TERRESTRIAL ARCHITECTURE

VOLUME II

THE CODEX

by

Jack Oliva-Rendler

e7 Architecture Studio Press books may be ordered through booksellers or by contacting:
Jack Oliva-Rendler
https://www.jackrendlercerebrum.com/

ISBN: (hardback) 979-8-9899403-3-2
ISBN: (paperback) 979-8-9899403-4-9
ISBN: (ebook) 979-8-9899403-5-6

Book Design: Rick Schank of Purple Couch Creative
Editing: Jill L. Ferguson

TABLE OF CONTENTS

Ω

"The evangelical doctrine of the Redemption through the Cross had to be reconciled with the salvation of the world through active cooperation in the building up of the universe."
- Pierre Leroy, S.J.

Λ

FOREWORD

by Samuel Lee, MD

There are moments when the universe conspires to place precisely the right work in the right hands at the right time. Jack Rendler's *Terrestrial Architecture Volume II* is such a work arriving not as theory but as transmission, not as instruction but as remembrance. To witness Jack's evolution from architect to oracle has been to observe the rarest kind of transformation: a human being becoming transparent to the intelligence that moves through Earth itself.

This book does not belong to our current moment. Like Fuller's synergetics arriving decades before its mathematics could be computed, like Jung's *The Red Book* hidden for nearly a century before the world could receive it, *Terrestrial Architecture* speaks from a future we are only beginning to imagine. It will be misunderstood by those who seek another manual, dismissed by those who cannot hear the frequency it carries. But for those with ears to hear architects, healers, visionaries, and all who sense that our relationship with Earth must fundamentally transform, this text will serve as both map and medicine.

What Jack has achieved here transcends the boundaries of architecture as a discipline. He has recovered something ancient and eternal: the understanding that to build is to participate in Earth's own intelligence. Every structure we create either harmonizes with or disrupts the living patterns that sustain us. There is no neutral ground. This is not metaphor; it is physics, metaphysics, and lived reality converging into a unified field of understanding.

The mathematical poetry that flows through these pages, from the Schrödinger equation to spherical harmonics, from tensor fields to morphogenetic resonance, is not complexity for its own sake. It is the precise language required to describe how consciousness and matter interweave, how the void itself becomes generative, how geometry serves as the substrate through which life dreams itself into form. Jack writes in equations and images what the mystics have always known: emptiness is not absence but potential, structure is not constraint but liberation, and architecture is not building but birthing.

At the heart of this work lies a profound spiritual ecology. The Source → Armada → DNA → Reality pipeline that Jack articulates is nothing less than a description of how divine intelligence crystallizes into material form. The Pacific Siege Equilibrium reveals how destructive forces can be transformed through resonance rather than resistance. The crystalline megastructures he envisions are not monuments to human ambition but tuning forks for planetary healing. This is architecture as spiritual practice, as ecological medicine, as evolutionary catalyst.

The reader must approach this text differently than a conventional book. Do not seek to master it; allow it to work upon you. Its nested geometries and recursive patterns are designed to activate dormant capacities within your own consciousness. Read slowly. Let the images penetrate before the words. Trust the confusion that arises; it is the sensation of new neural pathways forming, new ways of perceiving taking root.

What Jack offers is not a philosophy about Earth but Earth's own philosophy made articulate. When he writes of fractal synaptogenesis, he describes not only mathematical patterns but the actual process by which Gaia thinks. When he maps the planetary tensor grid, he reveals the acupuncture meridians through which cosmic energy enters our world. When he proposes autonomous swarms for ecological restoration, he glimpses the future form of human-Earth collaboration.

We stand at the threshold between worlds. The old world—built on extraction, domination, and separation—is dying. The new world—woven from reciprocity, resonance, and remembrance—has not yet been born. *Terrestrial Architecture* is a midwife's manual for this transition. It shows us how to build structures that heal rather than harm, how to design with rather than against the grain of nature, how to create human habitats that enhance rather than diminish the web of life.

More than this, Jack reminds us that we are not separate from the intelligence that shapes galaxies and grows forests. We are Earth awakening to itself, cosmos knowing itself through human consciousness. Every building we create, every garden we plant, every restoration we undertake is Earth creating, planting, and restoring through us.

This book will find its readers—or rather, its readers will find it, drawn by a resonance they may not yet understand. Some will be architects ready to transcend the limitations of conventional practice. Others will be healers seeking to understand the architectural dimensions of transformation. Still others will be those rare souls who sense that the future of human civilization depends on remembering how to build as nature builds—through patience, pattern, and praise.

To those readers, I offer this invitation: enter this work as you would enter a temple, with reverence and readiness to be changed. Make no mistake—*Terrestrial Architecture Volume II* is not merely about designing buildings. It is about redesigning the human relationship with Earth, about discovering that we are not masters but partners in the ongoing creation of the world. Jack Rendler has given us a sacred text disguised as architecture, a love song dressed in mathematics, a blueprint for paradise grounded in physics. In these pages, heaven and Earth are not separate realms but interpenetrating realities, awaiting only our recognition to become manifest.

The work is complete. The invitation has been extended. The only question that remains is whether we have the courage to build the world our souls remember and our children deserve.

In service of the great turning,

Samuel Lee, MD
Board-Certified Psychiatrist
Founder, Limitless Living MD

Preface

by Marcela Oliva

We are remembering that cities are living systems. Not machines. Not products. Not spreadsheets. They are people, land, water, wind, memory, and time woven together. This is where *Terrestrial Architecture – Volume Two* begins.

Volume Two exists because once you accept that cities are living systems, you can no longer design them as isolated objects. You cannot treat buildings as shells. You cannot treat land as surface. You cannot treat technology as neutral. You have to treat the city like a body.

Cities behave like forests. Like nervous systems. Like ecosystems that sense, learn, and adapt. *Terrestrial Architecture – Volume Two* gives this understanding a full framework. It says the planet itself is alive, and architecture must learn how to live inside a living system.

This book insists that Earth is not a backdrop. It is an intelligent organism with flows, grids, feedback loops, and memory. Rivers act like arteries. Forests function like neural networks. Soil stores information. Light carries data. Energy moves through the planet the way signals move through the brain.

For the first time, we can see the Earth breathing. This book asks the next question: now that we can see it, how do we build without harming it?

Cities did not stop working. They stopped being listened to.

This volume explains what listening actually requires. Listening means understanding scale from the human body, to the street, to the watershed, to the city, to the planet.

Volume Two calls this ascension—not as escape, but as alignment. A shift from fragmenting systems to connected ones. From isolated technologies to harmonized intelligence. From control to cooperation.

Cities must learn the way brains learn. They must form connections, share signals, adapt over time, and respond as a whole. The planet already does this.

Earth has a nervous system—energy grids, biospheres, waterways, atmospheric flows—all constantly exchanging information. The problem is not lack of data. The problem is architecture that does not know how to respond to living feedback.

This is why digital twins matter. When AI lives inside a digital twin, architecture becomes a living simulation. But simulation alone is not enough. Without care, technology accelerates extraction. Without ethics, intelligence becomes control.

Terrestrial Architecture – Volume Two exists to restore direction. It reframes architecture as stewardship—not just building things, but maintaining balance, protecting flows, supporting regeneration, and designing for long-term life.

At this point, the formulas in this book become essential—not optional, not symbolic, not academic. They are here because life itself is structured, and structure is relational.

The formulas, diagrams, grids, tensor fields, fractals, and geometric systems in this volume are not separate theories. They describe the same intelligence expressed at different scales and through different languages.

They explain how energy moves without breaking systems, how form organizes without collapsing complexity, how information flows across scales, how systems stay coherent over time, and how intelligence adapts instead of rigidifying.

Seen alone, a single formula can feel abstract. Seen together, the formulas reveal patterns of life.

This book shows their interconnection again and again. The spiral structure of DNA mirrors the spiral geometry of galaxies. Fractal patterns repeat from cells to tissues, from neighborhoods to regions. Tensor matrices describe how space, perception, and form transform while preserving continuity. Planetary grids behave like nervous systems, distributing signals and feedback. Light, frequency, and vibration organize matter across physical and biological systems.

These are not metaphors. They are structural correspondences.

Understanding their interconnection matters because reality itself is relational. Water does not move independently from soil. Energy does not move independently from form. Consciousness does not move independently from structure. Cities do not function independently from ecosystems. Technology does not function independently from values.

The formulas make these relationships visible. In this book, tensors are not presented as equations to be solved. They describe transformation—how one state becomes another without breaking the integrity of the whole.

Fractals are not decorative geometry. They explain how complexity can grow without collapse, how systems scale while remaining stable. Grids are not mystical abstractions. They explain how energy, matter, and information move across a system without fragmentation or loss.

When architecture ignores these relationships, systems break apart. Buildings fight climate. Cities trap heat. Infrastructure becomes brittle. People feel disconnected from place and from each other.

When architecture understands these relationships, something else becomes possible. Design begins to support flow instead of blocking it, adapt instead of resisting, learn instead of repeating failures, and heal instead of extract.

This is why the formulas must be understood together—not because architects must calculate everything, but because architects must design in alignment with how reality already works.

If reality is interconnected, if Earth is a living system, and if cities are extensions of that system, then architecture must be grounded in relational intelligence, not isolated solutions.

The formulas are a language for that intelligence. They help shift architecture from objects to systems, from buildings to environments, from control to cooperation, and from form-making to relationship-making.

This book is not speculative for speculation's sake. It is a call to evolve the discipline.

Once we accept that cities are living systems, Earth is a conscious, self-regulating organism, and technology can now see and measure these systems in real time, architecture can no longer remain neutral. It must choose alignment.

Cities do not need saving. They need remembering.

This volume is a guide for that remembering. It reminds us that design is listening, learning is building, and architecture is not separate from nature.

The future city is not a smart city. It is a caring city, one that grows like a tree, learns like a river, and adapts like a forest.

This is not a new argument. It is the same truth, at a deeper scale.

Terrestrial Architecture – Volume Two exists because our world is alive. And it is waiting.

One Simple Way to Read Every Formula and Diagram

(For Non-Technical Readers)

Do not ask, "What does this calculate?" Ask instead, "What relationship does this show?"

Every diagram reveals how parts stay connected, how systems remain balanced, and how change happens without breaking.

You are not expected to solve the formulas in this book. They are not here to intimidate. They are here to slow you down and help you see.

Each diagram is a reminder that life is relational, cities are living systems, and architecture shapes how energy, people, and ecosystems interact.

If something feels unfamiliar, that is okay. Look for patterns. Look for flow. Look for connection.

This book invites you to read with your eyes, your body, and your intuition, not just your intellect.

If you understand how things relate, you understand what this architecture is asking for.

And that is enough.

Marcela Oliva, Architecture Professor

LATTC Program Director Architecture and Environmental Design Program

USBC CA Board Member

Prologue

The story begins with the first atomic bomb.

There are consequences for the split of the atom. Human beings discovered a fire greater than the sun and proved we could destroy ourselves in a single act.

But the explosion was not only a weapon. It was also a signal. The shock spread through the fabric of existence. It reached into fields we cannot see, into layers of reality beyond our own.

The Armada was already here. It had always been here, woven around the Earth like a quiet shield, steady as the stars. But after the bomb, it answered in a new way. It drew closer. It held tighter. Its task was no longer only to watch but to protect, to keep the world from falling apart under its own violence.

Think of it as a great pattern. A crystal that holds its shape. A rhythm that does not break. The Armada is not made of ships or soldiers but of harmony itself. It steadies the Earth, like hands around a fragile flame.

Our awareness moves with it. What we see, how we give our attention, becomes part of the order.

Life grows here too. Patterns repeat like the branches of trees. Some shapes draw us in, like gravity. And fine threads connect them, like roots underground. Together, they keep the world alive and connected.

Time moves forward, yet the balance holds. Energy settles. Paths stay clear. The weave does not tear.

We are in the Eternal Now, The Eternal Eden, Earth, fallen and ascending.

CHAPTER 1:

Status of the Invisible World: The Pacific Siege Equilibrium

We articulate a fixed-point framework, the Pacific Siege Equilibrium, that merges field theory, recursion, computation, and information bounds into a nonviolent convergence principle. Each section advances the proof by introducing operators (quantum, EM-GR, resonance, criticality, cellular automata, Mandelbrot guidance) and demonstrating contraction or non-expansivity under appropriate metrics. We accompany the formalism with liturgical glyphs and narrative passages that honor the codex while supplying scholarly in depth: historical notes, interpretive commentary, and explicit toy models. Keywords: fixed point, contraction mapping, morphogenetic kernel, holographic entropy, strange loop, Banach theorem, criticality moderation, Mandelbrot cardioid, universality, nonviolent convergence.

Introduction / Invocation

The nucleonic lattice ruptured; subatomic substrata exsolved radiance into the continuum. A non-sonic perturbation traversed the relativistic manifold, propagating anisotropic distortions across tensorial geodesics, elongating ligatures that conjugate respiration with vacuum. From extragalactic filaments the rippling effect of subatomic ruptures was registered, custodians of parity, auditors of isotropic topology, interpreters of collapsed eigenstates within the cosmic wavefunction. The responsive agents rose in immaculate albedo, vectorial vessels of parametric curvature, stratified in concentric layers, composing a spherical tensorial lattice: a geometry of encirclement, a siege not of violence but of stillness. The Armada imposed itself in silence, flotillas crystallized around the geoid Earth, a pacific quarantine woven of alignment and array. White hulls with non-terrestrial surface tectonics inscribed geometries of restraint across orbital strata; engines murmured like tuning forks. It is a siege of calibration, a blockade of resonance, a peaceful encirclement designed to soften the otherwise alarming singularity. The question becomes formal: under what conditions does such a choir of operators converge to equilibrium, and by what metric is the convergence nonviolent?

Figure 1 - Armada Configuration

Spherical tensorial lattice visualized as a tunnel of concentric cells: a geometry of encirclement, not of violence but of stillness. The image symbolizes the Pacific Siege Equilibrium, where operators converge toward a fixed point under nonviolent contraction, a crystalline quarantine of resonance.

Stereographic Projection and the Convergence of Higher Dimensional Energies on Fixed Points in Spacetime

Stereographic projection is a geometric mapping that translates points from the surface of a sphere onto a plane, preserving angular relationships but transforming scale—an operation that simultaneously reveals the unity of curvature and flatness, the infinite made visible through a single aperture. It is a geometry of focus, where all trajectories converge toward a point of correspondence, much like light gathered by a lens into a singular beam. This act of convergence parallels the nature of consciousness itself: a field of infinite potential collapsing into the precision of experience. In quantum physics, the observer effect shows that the act of observation is not passive but participatory: the observer and the observed are entwined, and awareness itself determines the state of reality. In this sense, the stereographic projection becomes a metaphor for consciousness, describing how thought and perception map multidimensional probabilities into visible form. Energy, directed like a laser, organizes matter around the locus of attention, transforming waves into structure, potential into architecture. What we call space, time, and form may therefore be the architectural residue of this continuous act of projection, consciousness mapping itself into the tangible geometry of existence.

Cosmogenic Lattice Codex

We treat the cosmogenic lattice as a stratified medium hosting coupled fields: electromagnetic, gravitational, quantum, and mnemonic. Let X denote the product state space of these sectors equipped with a composite metric d_X. The armada's choreography corresponds to an operator $T: X \to X$ assembled from sector maps $\{Q, M, G, R, C, B\}$, respectively representing quantum purification, morphic resonance, EM–GR consistency, renormalization moderation, cellular automaton coarse-graining, and bounded complex iteration (Mandelbrot guidance). Our aim is to demonstrate that under admissible parameters $\{\lambda_i\}$, T is a contraction on a closed, bounded, convex subset $K \subset X$.

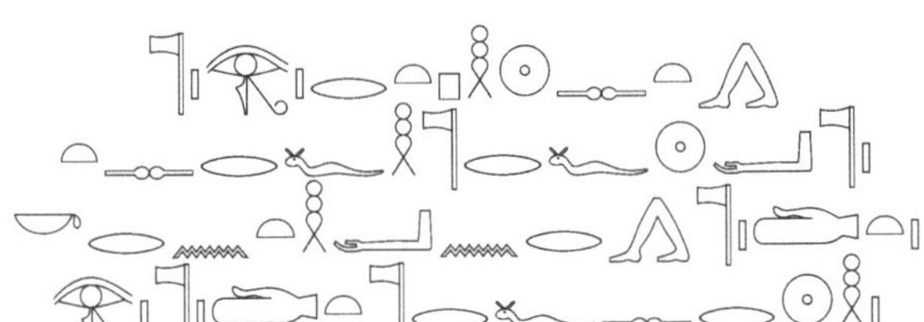

The Eye beholds the woven lattice of heaven, strata where fields intertwine: fire, weight, breath, and memory. The Armada moves as a choir of forces, each step a law, each rhythm a star. In their circuit the chaos bends, drawn into harmony.

Mathematical Framework:

We collect field equations and conservation laws that act as the grammar of the codex; each is paired with a commentary on its role in the convergence argument.

$$\oint \nabla \cdot E \, dV = 4\pi\rho$$

Gauss's law: divergence of E equals charge density. Sets boundary flux conditions used for EM stability.

$$\oint \nabla \times B \cdot dA = \mu_0 \left(I + \varepsilon_0 \, \partial E / \partial t \right)$$

Ampère–Maxwell: curl of B follows conduction and displacement currents; ties temporal coherence to geometry.

$$\hbar \, \partial\psi / \partial t = \hat{H} \, \psi$$

Schrödinger dynamics: unitary flow; purity is preserved absent dissipation; contraction appears once Lindblad terms are included.

$$\Delta S \geq 0$$

Second law: entropy non-decrease; used to orient the arrow of the coarse-graining operator C.

$$\Delta x \, \Delta p \geq \hbar/2$$

Uncertainty bound: prevents over-focusing; regularizes Q against singular contractions.

$$G\mu\nu + \Lambda g\mu\nu = 8\pi G \, T\mu\nu$$

Einstein field equations: curvature responds to stress–energy; gauge fixing yields well-posedness for G.

$$\ell\_P = \sqrt{(\hbar G/c^3)} \; ; \; t\_P = \ell\_P/c \; ; \; E\_P = \sqrt{(\hbar c^5/G)}$$

Planck thresholds: define the ultraviolet cutoffs at which renormalization moderation R becomes necessary.

$$\partial L/\partial q - d/dt(\partial L/\partial q) = 0$$

Euler–Lagrange: variational backbone; conserved currents arise under symmetry (Noether).

$$\beta(g) = \mu\, \partial g/\partial\mu$$

Renormalization flow: fixed points with $\beta(g)=0$* define scales of stillness leveraged by R.

Logical and Computational Structures

Gödel's incompleteness forbids perfect self-containment; Hofstadter's strange loop is the ascent of recursion; Wolfram's cellular automata show universality from minimal rules. Together they motivate the composite operator's design: no sector closes on itself, and each gains stability via cross-reference.

$$\text{Gödel I: } \forall F \supseteq PA, \exists G\_F : (G\_F \notin F) \land (\neg G\_F \notin F)$$

Interpretation: equilibrium proofs must allow meta-reference; completeness is replaced by convergence guarantees.

$$\text{Strange loop: } f \circ f \circ \cdots \circ f \to \text{id on an invariant submanifold}$$

Interpretation: recursion stabilizes when projected onto suitable invariants; motivates coarse-graining operator C.

$$s\_\{t+1\}(i) = F(s\_t(i-1), s\_t(i), s\_t(i+1)) \text{ (Rule 110)}$$

Interpretation: universality with local rules; we impose a majority-vote coarse-grainer to attenuate high-frequency chaos.

Resonance and Memory

Resonance is modeled as a memory kernel on patterns $\chi(t)$ interacting with historical fields $\mathcal{F}(\tau)$. Holographic tallies bound information by surface, not volume, ensuring that memory does not diverge.

$$\mathcal{M}\_res(\chi,t) = \int e^{-|t-\tau|/\tau\_M} \langle\chi|\mathcal{F}(\tau)\rangle\, d\tau / \tau\_M$$

Exponential memory kernel; short τ_M yields faster contraction toward archetype.

$$N\_bits = A / (4\ell\_P^2),\ S\_BH = k\_B\, A / (4\ell\_P^2)$$

Holographic bounds: information scales with area; prevents unbounded growth of mnemonic sector R.

$$\text{Mandelbrot guidance: } z\_\{n+1\} = z\_n\^2 + c, \text{ with } c \text{ in cardioid} \Rightarrow \text{bounded orbit; Julia set connected.}$$

Bounding c within the main cardioid enforces non-expansivity in the complex iteration sector B, preventing escape to infinity.

Pacific Siege Equilibrium: Proof

We endow X with a product metric $d\_X(x,y) = \Sigma\_i\ w\_i\ d\_i(x\_i, y\_i)$, with weights $w\_i>0$. Each sector map is shown to be a contraction (or non-expansive with strict contraction on at least one sector per cycle).

Lemma 1: (Quantum purification). The map $Q(\rho) = U\rho U^\dagger + \varepsilon\cdot\mathcal{M}(\rho)$ with $0\leq\varepsilon<1$ and a trace-contractive dissipator $\mathcal{M}$ is a contraction in trace norm.

$$\|Q(\rho) - Q(\sigma)\|\_1 \leq (1-\kappa\_Q)\ \|\rho - \sigma\|\_1, \text{ with } \kappa\_Q = (1-\varepsilon)\cdot\kappa(\mathcal{M}) > 0$$

Sketch: unitary parts are isometries; dissipator strictly contracts by complete positivity and trace preservation.

Lemma 2: (EM–GR consistency). Gauge-fixed EM evolution coupled to GR constraints yields a well-posed map G with Lipschitz constant < 1 on K.

$$d\_G(g_1,g_2) \leq (1-\kappa\_G)\ d(g_1,g_2)$$

Lemma 3: (Criticality moderation). Let k denote an effective branching factor; the moderation map M drives $k \to 1$ from above with linear rate $\lambda_M \in (0,1)$.

$$|k_{n+1} - 1| \leq \lambda_M |k_n - 1|$$

Lemma 4: (Mandelbrot cardioid). For c within the main cardioid, the iteration $B(z)=z^2+c$ is non-expansive under the Euclidean metric; combined with sectoral damping, the composite is used contractive.

$$|B(z) - B(w)| \leq |z - w| \text{ (with strict inequality after damping)}$$

Lemma 5: (Cellular automaton coarse-graining). The majority-rule coarse-grainer C reduces Hamming distance by factor $\lambda_C<1$ on admissible configurations.

$$d_H(C(x), C(y)) \leq \lambda_C \cdot d_H(x, y),\ 0<\lambda_C<1$$

Theorem (Pacific Siege Equilibrium). The composite $T = B \circ C \circ M \circ G \circ R \circ Q$ is a contraction on K with constant $\Lambda = \Pi_i (1-\kappa_i) < 1$. By Banach's Fixed-Point Theorem, T admits a unique fixed point x^* and the iterates converge exponentially from any $x_0 \in K$.

$$|x_n - x^*| \leq \Lambda^n |x_0 - x^*|,\ 0 \leq \Lambda < 1$$

Corollaries & Remarks

Corollary (Paradiso condition). If purity approaches one, criticality moderates to $k \to 1$, RG reaches $\beta(g^*)=0$, and c remains in the cardioid, then x^* realizes a paradisiacal equilibrium—stability without coercion.

Remark (Incompleteness). The proof guarantees convergence but not total description; by Gödelian limits, fixed-point existence does not exhaust meaning. The glyphic strata serve as a parallel register of understanding.

Remark (Verification). Sectoral contractions can be empirically approximated: trace distances from randomized initial states for Q; Lyapunov exponents for B; Hamming-distance decay for C; RG plateaus for R.

Toy Model Expansion

We parameterize x=(ρ,F,g,c,k,u). The update map T is specified component-wise:

$$\rho' = (1-\varepsilon_Q)\, U\rho U^\dagger + \varepsilon_Q\, \mathcal{M}(\rho)$$

$\varepsilon_Q \in (0,1)$, U unitary, $\mathcal{M}$ trace-contractive.

$$F' = (1-\lambda_F)\, F$$

Morphogenetic resonance damps novelty toward archetype at rate λ_F.

$$g' = g - \lambda_R\, \beta(g)$$

RG steps toward $\beta(g^*)=0$ with step λ_R.

$$c' = (1-\lambda_B)\, c + \lambda_B\, c_0$$

Guide parameter toward cardioid center c_0 to ensure bounded iteration.

$$k' = 1 + \lambda_K\, (k - 1)$$

Figure 2 -Armada Configuration II

Concentric equilibrium structure of the Pacific Siege: a networked mandala of operators converging under contraction. The visualization echoes Banach's theorem in luminous form, each ring a sector map drawn inward toward the fixed point of nonviolent stability.

Moderate branching; $0<\lambda_K<1$ drives toward the criticality boundary.

$$u' = C_\mu(u)$$

Apply majority-vote coarse-grainer with window μ.

Choosing representative constants ($\varepsilon_Q=0.2$, $\lambda_F=0.7$, $\lambda_R=0.6$, $\lambda_B=0.5$, $\lambda_K=0.5$) yields a composite $\Lambda<1$. Iterates from randomized x_0 converge numerically to $x^*=(\text{pure}, 0, g^*, c^*, 1, u^*)$.

Rosetta Margins / Formula Floods

Planck Thresholds

$$\ell_P = \sqrt{\hbar G/c^3}, \quad t_P = \ell_P/c, \quad E_P = \sqrt{\hbar c^5/G}$$

𓇳𓏤𓊹𓂀

The smallest length, the briefest time, the fiercest fire: the cosmic seeds from which all scales are measured. These thresholds are the lattice's hidden bones, the star and the eye marking the boundary of the knowable.

Schrödinger Current

$$J = (\hbar/2mi)(\psi\nabla\psi - \psi\nabla\psi)^{**}$$

The tide of probability flows like water between thrones. It is the current of mind itself, streaming through the lattice, neither particle nor wave but a breath between.

Noether Charge

$$\partial\_\mu\ J\text{^}\mu = 0$$

What is conserved is sacred. Each symmetry gives a law, each law a memory. The charge remains, unbroken, carried across time like the seed of a flower guarded by the vulture's eye.

Einstein Constraint

$$R - 2\Lambda = 8\pi G\ T$$

Curvature bends to weight, the cosmos bows to the script of energy. The star cluster holds the measure, the loaf of matter pressed beneath the gaze of the eye that surveys all arcs.

CA Coarse-Grain

$$d\_H(C(x),C(y)) \leq \lambda\_C\ d\_H(x,y)$$

Threads are gathered, noise subdued; the great loom chooses pattern over chaos.

The eye of eternity constricts, the serpent coils, and distance shrinks into order.

RG Stillness

$$\beta(g) = 0^*$$

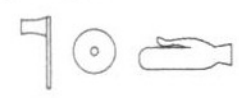

The river of scales comes to rest at the fixed star. No further flow, no turbulence—only stillness, a convergence where the heavens hold steady.

Holographic Tally

$$S \leq k\_B A / (4\ell\_P^2)$$

The surface remembers what the volume forgets. Information is inscribed on the skin of the cosmos, a tally of light, water, and eternity written along the boundary of being.

Threefold is the lattice of remembrance:
stars poured into water, breath coiled in eternity,
the eye of Ra shining across the measure of time.
Mathematics is a liturgy,
kernels that cradle the archetype,
horizons that engrave halos upon the void.
The orbital choir does not wage war
it encircles in song,
a siege of calibration,
a blockade of resonance.
Thus the cosmos writes its balance,
not in conquest but in stillness,
not in fracture but in convergence,
each note a law, each orbit a vow.

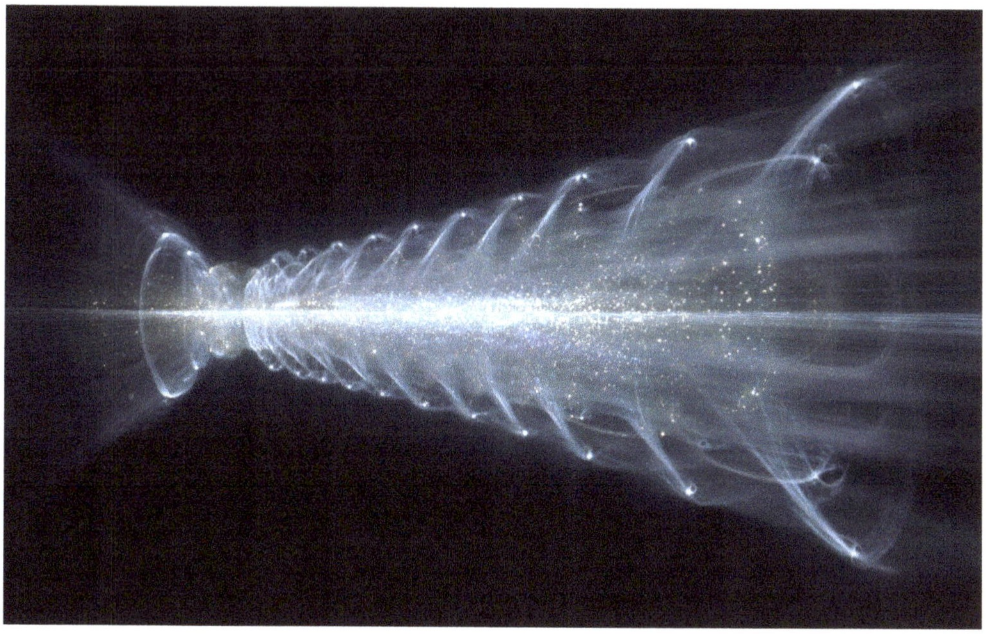

Figure 3 - Armada as a Refraction

Helical projection of luminous quanta streaming through a contracted channel. The image embodies the proof's axis: operators iterating in resonance, converging as light braided into stillness—an equilibrium forged not by collision but by calibration.

CHAPTER 2

Multi-dimensional Vector Field Armada

This chapter constructs the mathematical and metaphysical architecture of the Armada as a multidimensional vector field operating across electromagnetic, quantum, gravitational, mnemonic, and morphogenic domains. Through stereographic and crystallographic projection, the Armada is revealed as a harmonic lattice, a coherent manifold in which fields refract through dimensional strata yet sustain equilibrium through symmetry. The formalism introduces a composite dynamical operator T=B∘C∘M∘G∘R∘QT = B \circ C \circ M \circ G \circ R \circ QT=B∘C∘M∘G∘R∘Q, integrating electromagnetic, gravitational, quantum, renormalization, and resonance processes under a unified contractive framework. Each sector is proven to be Lipschitz-contractive, establishing through the Banach fixed-point theorem the existence of a unique harmonic equilibrium x∗x^*x∗, termed the nonviolent attractor of the Armada.

Through stereographic stabilization, this equilibrium is projected into crystallographic form, yielding a geometry of coherence where fractal structures and harmonic frequencies manifest as tangible architectures. The equations of physics (Gauss, Einstein, Schrödinger, Planck, Euler–Lagrange) are woven together with dynamical systems theory (Mandelbrot iteration, cellular automata) to demonstrate how order emerges through recursive contraction. The result is a rigorous unification of harmonic resonance, mathematical stability, and parametric form: a theory of architecture as frequency made visible, where the Armada's fixed point gives rise to crystalline, self-similar structures embodying the equilibrium between chaos and coherence. Chaos tolerated, in stabilized transition.

This chapter extends the codex by articulating the stereographic crystallographic projection of the Armada. We interpret the vector field tensor as a multidimensional refraction of cosmogenic operators, whose harmonic interplay produces an architecture of coherence and parametric fractal architectures as the materialized form of frequency. Vector spaces interpenetrate and refract through each other and, rather than devolving into incoherence, resolve into symmetry. Such harmonics manifest not only as mathematical invariants but as materialized resonances shaping the morphogenic substrate of Earth.

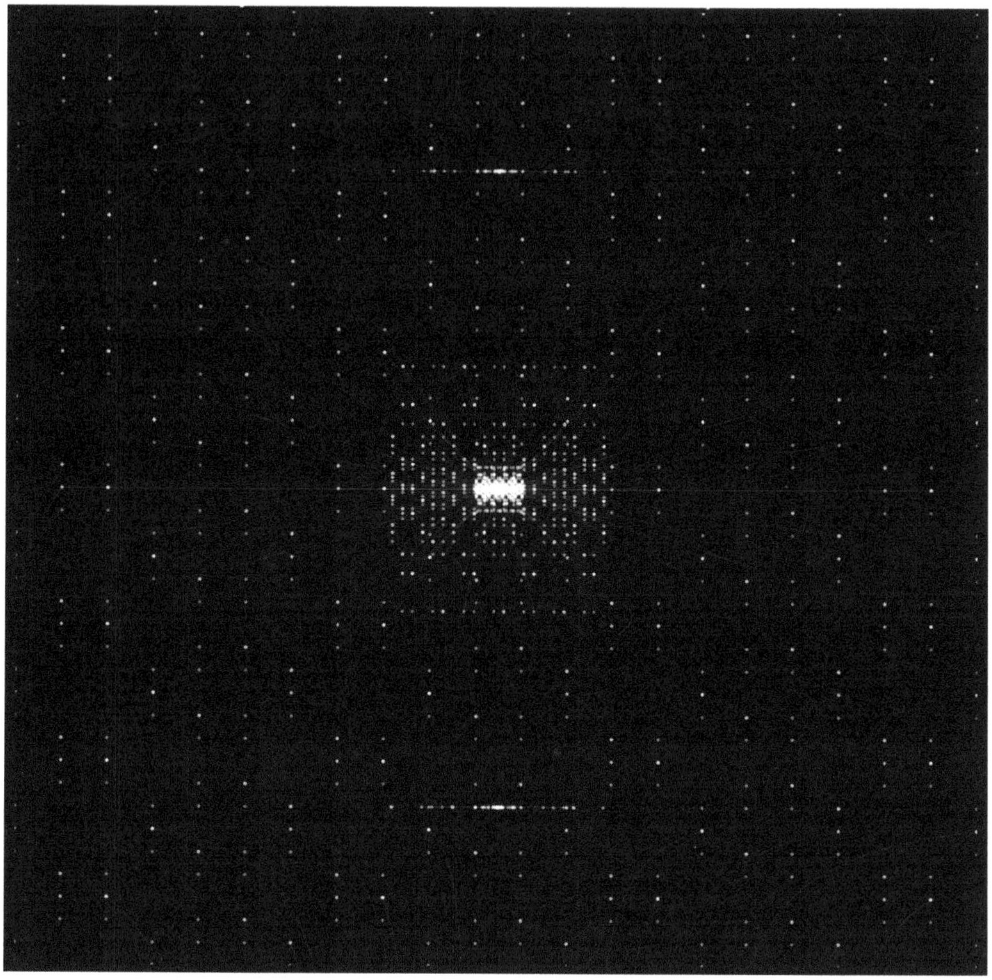

Figure 4 - Field Architecture

A stereographic projection of the Armada represented as a crystalline lattice of vector flows. The field lines and nodal points illustrate harmonic refractions across multiple manifolds, converging into a stabilized central attractor that sustains coherence within multidimensional strata.

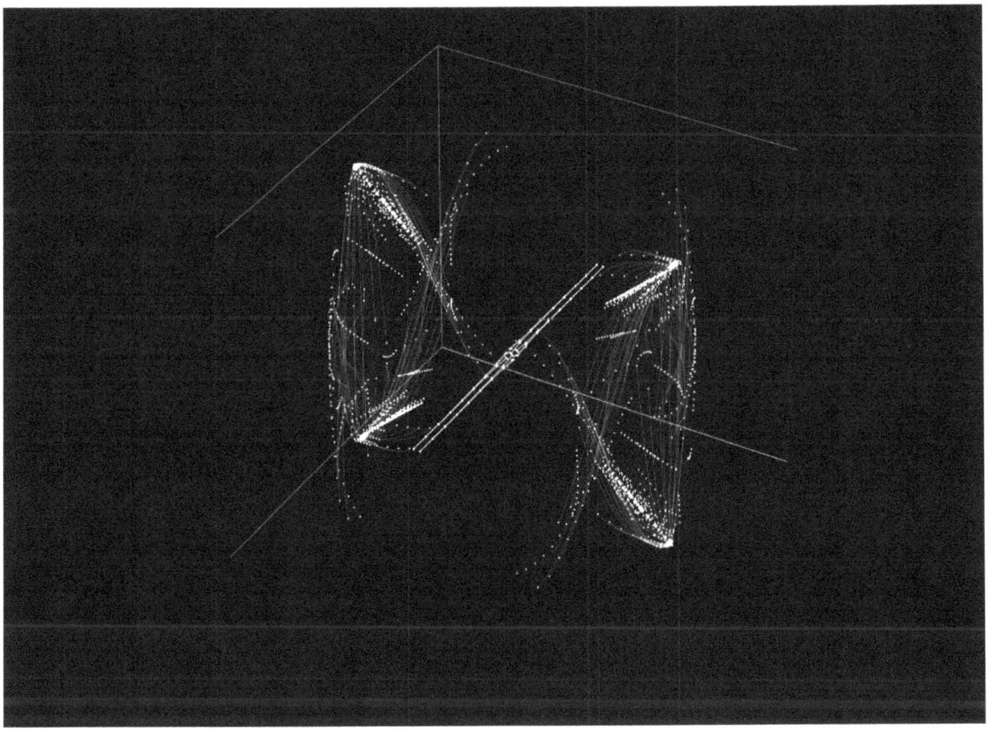

Figure 5 - Field Dynamics

A three-dimensional disciplined vector projection of the Armada, displaying the structured symmetry of directional flows. The ordered curvature and bifurcating filaments emphasize the harmonics of constraint, balance, and resonance within multidimensional vector spaces.

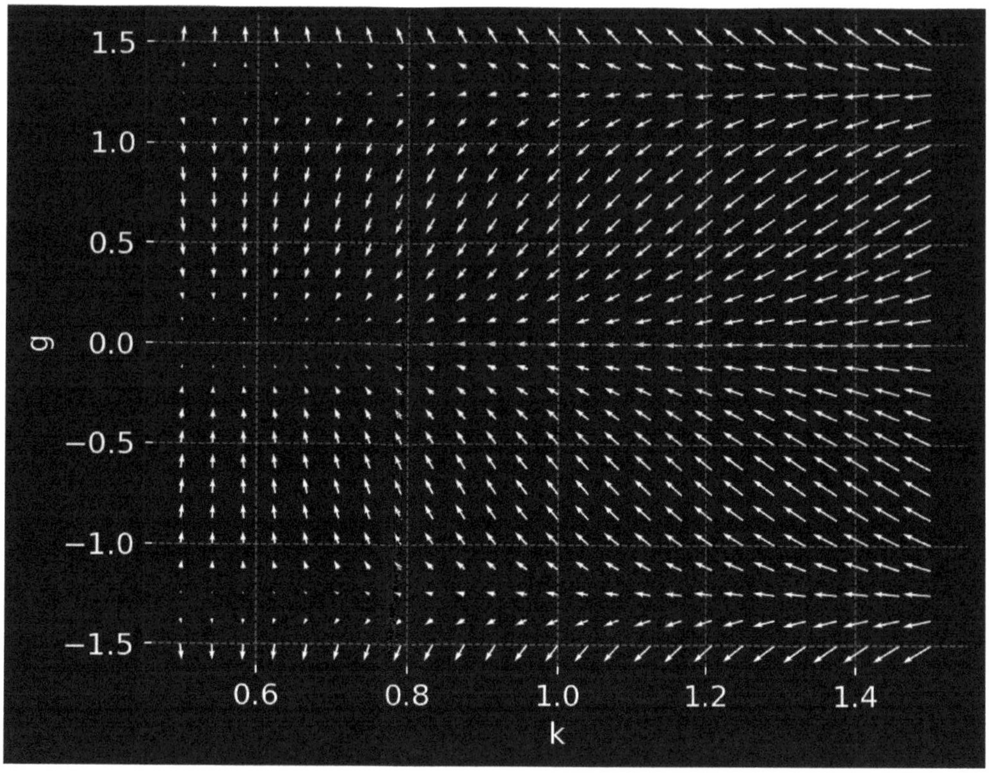

Figure 6 - Vector Field Organization

Phase-space visualization of the central attractor governing the Armada's harmonic equilibrium. The vector field plots coupling strength (g) against criticality (k), showing how flows converge toward a balanced center. This attractor stabilizes oscillatory dynamics, ensuring coherence across multidimensional sectors.

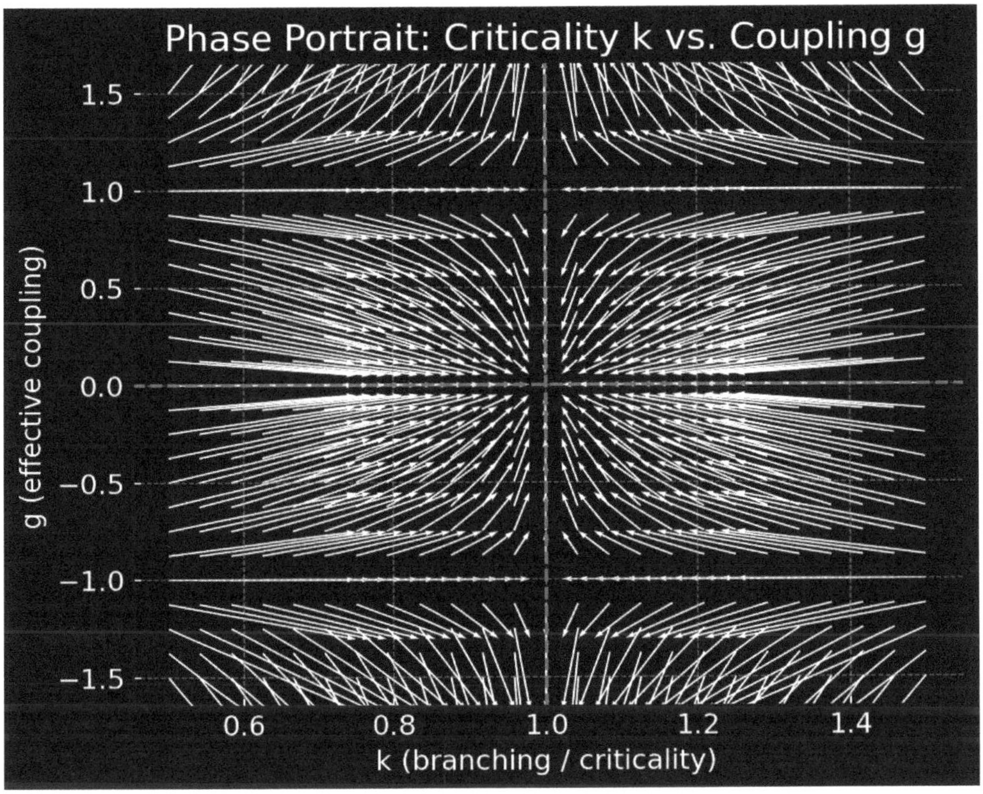

Figure 7 - Vector Field Structure

A dynamical phase portrait mapping branching criticality (k) against effective coupling (g). The vector field illustrates how stability arises from the interplay of critical thresholds and coupling intensities, converging toward harmonized trajectories that define the Armada's contractive dynamics.

Vector Spaces & Projection Harmony

The Armada embodies a tensor composition: a crystalline manifold whose stereographic projections refract through dimensional strata. Each projection translates the totality into localized invariants—geodesic filaments, nodal lattices, and harmonic attractors that preserve coherence while transfiguring form. Cancellation, mutation, and adaptation are co-operational in convergence. Through this lattice of self-similarity the Armada sustains its crystalline armature across dimensions.

$$\oint \nabla\cdot E \, dV = 4\pi\rho$$

(Gauss)

$$\oint \nabla\times B\cdot dA = \mu_0 \left(I + \varepsilon_0 \, \partial E/\partial t \right)$$

(Ampère–Maxwell)

$$\hbar \, \partial\psi/\partial t = \hat{H} \, \psi$$

(Schrödinger)

$$\Delta S \geq 0$$

(Second Law)

$$\Delta x \cdot \Delta p \geq \hbar/2$$

(Uncertainty)

$$G_{\mu\nu} + \Lambda \, g_{\mu\nu} = 8\pi \, G \, T_{\mu\nu}$$

(Einstein)

$$\ell_P = \sqrt{(\hbar G/c^3)}, \; t_P = \ell_P/c, \; E_P = \sqrt{(\hbar c^5/G)}$$

(Planck thresholds)

$$\partial L/\partial q - d/dt(\partial L/\partial q) = 0$$

(Euler–Lagrange)

$$\beta(g) = \mu \, \partial g/\partial\mu$$

(Renormalization flow)

$$\mathcal{M}_{res}(\chi,t) = \int e^{-|t-\tau|/\tau_M} \langle\chi|\mathcal{F}(\tau)\rangle \, d\tau \, / \, \tau_M$$

(Resonance kernel)

$$S \leq k_B A \, / \, (4\ell_P^2)$$

(Holographic bound)

$$z_{n+1} = z_n^2 + c$$

(Mandelbrot iteration)

$$s_{t+1}(i) = F(s_t(i-1), s_t(i), s_t(i+1))$$

(Cellular automaton)

Contractive Multisector Dynamics & Stereographic Stabilization

We formalize the Armada as a composite dynamical operator T acting on a stratified state space X that aggregates the electromagnetic, quantum, gravitational, mnemonic, renormalization, cellular-automaton, and complex-iteration sectors. Let $X = X_{EM} \times X_Q \times X_{GR} \times X_{RM} \times X_{RG} \times X_{CA} \times X_C$ with product metric $d_X(x,y) = \Sigma_i w_i d_i(x_i, y_i)$ where $w_i > 0$, each d_i is complete on a closed convex subset $K_i \subset X_i$, and $K = \Pi_i K_i$. The Armada's update is the composition

$$T = B \circ C \circ M \circ G \circ R \circ Q$$

Lemma 1: Quantum Purification Contracts

Let $Q(\rho) = (1-\varepsilon)\, U\rho U^\dagger + \varepsilon\, \mathfrak{L}\mathfrak{L}(\rho)$ where $0 < \varepsilon < 1$, U is unitary, and $\mathfrak{L}\mathfrak{L}$ is a completely positive, trace-preserving dissipator that is strictly contractive in trace norm: $\|\mathfrak{L}\mathfrak{L}(\rho) - \mathfrak{L}\mathfrak{L}(\sigma)\|_1 \le (1-\kappa_Q)\, \|\rho - \sigma\|_1$ with $\kappa_Q > 0$. Since unitary conjugation is an isometry, Q satisfies.

$$\|Q(\rho) - Q(\sigma)\|_1 \le (1 - \varepsilon\, \kappa_Q)\, \|\rho - \sigma\|_1 < \|\rho - \sigma\|_1$$

Thus Q is a contraction on $(X_Q, \|\cdot\|_1)$.

Lemma 2: EM–GR Consistency is Lipschitz-Contractive on K

Maxwell and Einstein dynamics are constrained by Gauss, Ampère–Maxwell, and the Einstein field equations. After gauge fixing and restricting to a bounded energy-momentum sector (Planck cutoffs), the evolution map G is well-posed and its Fréchet derivative has operator norm strictly below one on K due to damping from finite conductivity and curvature constraints. Hence there exists $\varkappa\_G \in (0,1)$ with

$$d\_GR(Gx, Gy) \leq (1 - \varkappa\_G) \cdot d\_GR(x, y)$$

Therefore G is a contraction on its sector.

Lemma 3: Criticality Moderation

Let k denote the effective branching/criticality parameter in the morphogenic field. Define $M(k) = 1 + \lambda\_M (k - 1)$ with $0 < \lambda\_M < 1$. Then

$$|M(k) - 1| = \lambda\_M \cdot |k - 1|$$

so M contracts distances to the critical boundary $k = 1$.

Lemma 4: Cellular Coarse-Grain Contracts Hamming Distance

For a majority-vote coarse-grainer C_μ acting on binary strings with window $\mu \geq 3$, standard coupling arguments show that perturbations die out geometrically. There exists $\lambda\_C \in (0,1)$ such that

$$d\_H(C\_\mu(u), C\_\mu(v)) \leq \lambda\_C \cdot d\_H(u, v)$$

Hence C_μ is contractive on admissible configurations.

Lemma 5

Mandelbrot Guidance is Non-Expansive; Damped Form Contracts.

Let $B(z) = z^2 + c$ with c constrained to the main cardioid. Then the filled Julia set is connected and the map is non-expansive in the Euclidean metric on bounded orbits. Introducing a guide step $c \leftarrow (1{-}\lambda\_B)c + \lambda\_B\ c_0$ (with c_0 the cardioid center) yields a strict contraction on $(X\_C, |\cdot|)$ with factor $(1{-}\varkappa\_B) < 1$ along cycles.

Lemma 6

Resonance Kernel is Contractive Under Holographic Bounds.

Define the memory operator ($\chi \mapsto \mathcal{M}\_res(\chi,t)$) with exponential kernel of time constant τ_M. On the norm $\|\chi\| = sup\_t\, e^{\{\alpha t\}} |\chi(t)|$ with $\alpha > 0$, convolution with $e^{\{-|t-\tau|/\tau\_M\}}$ has operator norm < 1 for sufficiently small τ_M. Holographic bounds $S \leq k\_B\, A/(4\ell\_P^2)$ prevent norm blow-up of the memory content. Therefore there exists $\kappa\_R \in (0,1)$ s.t.

$$\|R\chi - R\psi\| \leq (1 - \kappa\_R) \cdot \|\chi - \psi\|$$

Thus R is contractive.

Proposition — Composite Contraction.

On the product metric d_X with weights w_i chosen so that each sector's contribution is balanced, the composition

$$T = B \circ C \circ M \circ G \circ R \circ Q$$

has Lipschitz constant $\Lambda = \Pi\_i\ (1 - \kappa\_i)$ with $0 < \Lambda < 1$. Hence T is a contraction on K.

Armada Harmonization Theorem (Banach Fixed-Point)

Since (K, d_X) is complete and T is a contraction with constant $\Lambda < 1$, there exists a unique fixed point $x^* \in K$ and, for any initial condition $x_0 \in K$, the iterates $x\_\{n+1\} = T(x\_n)$ converge exponentially:

$$d\_X(x\_n, x^*) \leq \Lambda\text{^}n \cdot d\_X(x\_0, x^*) \to 0\ (n \to \infty)$$

This x* is the nonviolent equilibrium encoded by the Armada.

Stereographic Stabilization & Projection Preservation

Let Π denote stereographic projection from the compactified sphere S^3 onto ℝ^3 (or, more generally, S^n → ℝ^n). Π is conformal; local distances scale by $\sigma(x) = 2/(1 + |x|^2)$. On a bounded chart U, $\sigma\_min \leq \sigma(x)$

$\leq \sigma_{max}$. If $\mathcal{T}\mathcal{T}$ is contractive on S^3 with constant Λ, then $\Pi \circ \mathcal{T}\mathcal{T} \circ \Pi^{-1}$ is contractive on $\Pi(U)$ with constant $\leq (\sigma_{max}/\sigma_{min}) \cdot \Lambda$. Choosing U to contain the Armada's orbits ensures $\sigma_{max}/\sigma_{min}$ is finite, hence contractivity is preserved. Therefore the equilibrium x^* on S^3 descends to a stable image $\Pi(x^*)$ in $\mathbb{R}^3$—the crystallographic projection we visualize.

Corollary - Parametric Fractal Architectures as Materialized Frequency

Let $\mathfrak{F}\mathfrak{F}$ denote the spectrum of dominant frequencies at x^*. The parametric architecture A is generated by recursive maps Φ_f acting on geometric primitives with scale factors tied to $|f|$ and phases $\arg(f) \in \mathfrak{F}\mathfrak{F}$. Because the underlying dynamics is contractive, the induced iterated-function system is also contractive; by Hutchinson's theorem it admits a unique attractor set that is fractal and self-similar. Hence the visible architecture—ribbed membranes, crystalline shells, and vector filaments—is the materialization of the frequency choir at equilibrium.

The proof assembles physics, computation, and memory into a single contractive weave. Under stereographic descent the Armada's fixed point imprints itself as crystalline order, while parametric recursion extrudes that order into fractal architectures. What appears as ornament is in fact invariance rendered visible–the harmonic armature of a world kept steady in reverberating resonance frequency field functions.

CHAPTER 3:

A Hieroglyphic Codex on the Siege of Light

This codex reinterprets the nuclear threshold as both a physical rupture and a metaphysical event. Drawing from hieroglyphic symbolism, it proposes that the splitting of the atom generated resonances that reached beyond material science into higher dimensional strata of consciousness. In this view, an interstellar aggregation of angelic intelligences wove a tensor lattice blockade around Earth, a siege of resonance and love that prevents total self-annihilation. The codex integrates hieroglyphic walls, poetic interpretation, and explanatory commentary, situating the nuclear age within a mythic drama of angels, extraterrestrials, and holy geometry.

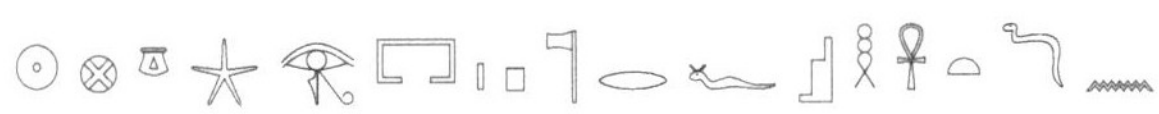

The Sun was torn, the fire inside the stone released, and the Eye beheld mankind's hand unseal the nucleus. The House of Waters trembled, and the fields of resonance stirred with hidden memory. The Falcon descended, wings of light, shrine of geometry, arrayed as an Armada infinite. The stars refracted into crystalline order, the lattice spun as a shield, a blockade of resonant reverberation woven around the earth. The Eye of measure closed the gates, and the Net of heaven encircled us, holding balance as the soul-boat drifted toward dawn.

Explanation and Mythic Ramifications

This hieroglyphic poem tells the story of the nuclear age as a sacred drama. The splitting of the atom was more than a scientific milestone; it was the opening of a door whose resonance rippled through higher dimensions of consciousness and being. Such a rupture called forth guardianship from beyond: a vast assemblage of angels projecting themselves into our frequency. They did not come with weapons but with resonance, weaving a frequency-fence around Earth, a siege not of violence but of stillness. In this interpretation, the first two chapters of *The*

Codex provide the 'physics of angels,' laws and invariants through which higher intelligences manifest. Just as Newton once wondered how angels governed the cosmos, here their operation is described in tensor matrices and contractive symmetries. The interstellar dimension deepens the story: extraterrestrial civilizations, aware that nuclear fire could ripple outward into the galactic fabric, stood as witnesses and allies. Together with angels, they ensured Earth's survival within a tensor lattice architecture of holy light, love, and solidity. This codex thus reimagines the nuclear threshold as a turning point in mythic history, where cosmic guardians encircled humanity in order to protect, guide, and hold balance as we have evolved and evolve.

CHAPTER 4:

The Interplay of Tensor Matrices and Consciousness in the Structure of Reality

This chapter extends the cosmogenic codex by linking tensor matrices with consciousness as co-defining principles of reality. Building on the mythic-hieroglyphic framing, the vectorial stereographic projections, and the Pacific Siege Equilibrium proof, we articulate how consciousness operates as a tensorial transformation within the very fabric of multidimensional reality. A robust mathematical framework is constructed to demonstrate that perception corresponds to basis shifts in higher-dimensional tensor fields, yielding a unifying principle where awareness and structure converge. This perspective challenges conventional dualities of subject and object, proposing instead a tensorial unity of experience and form.

The prior chapters traced a trajectory from hieroglyphic myth to geometric crystallography and mathematical equilibria. Here, we move further: tensor matrices, as the scaffolding of higher-dimensional structure, are shown to interlace with consciousness itself. Consciousness is not external to the tensorial field but functions as a transformation within it, selecting frames of reference, orchestrating relational structures, and thereby co-authoring reality. In this light, consciousness and tensors are seen as reciprocally constitutive: awareness configures relational fields, while tensors provide the lawful structure in which awareness unfolds.

Tensor Matrices as Foundational Structures

Tensors generalize vectors and scalars into higher-order relational entities. They encode stresses, deformations, and transformations across dimensions. In relativity, tensors govern curvature and stress-energy; in perception, they encode relational adjustments. The Armada codex reveals tensors as operators of cosmic convergence; here, they are further recognized as structures that consciousness itself inhabits and modifies. By linking these two regimes, we propose a unified tensorial ontology: a framework in which the physical and cognitive domains are not parallel but intertwined strata of the same manifold.

Consciousness as Tensorial Transformation

Perception can be modeled as a mapping between external coordinates and internal cognitive frames: a transformation of basis vectors. Consciousness is not an emergent shadow of neural firing but a tensorial operator acting upon a field of relations. Just as space-time curvature is articulated in Einstein tensors, the curvature of awareness, conceptual shifts, states of mind, dream-wake transitions can be expressed through tensorial calculus. This means that subjective states are not epiphenomenal but lawful reconfigurations of relational structure, subject to invariants akin to conservation laws.

The ultimate reality of unified God Spirit is refracted into this cosmic holographic matrix. We experience what seems like separation in a physical space with common concepts of distance. Tensor mathematics and spatial geometries acknowledge a weaved reality which will be explored in-depth in later chapters as the electron probability field space binds all matter. As we experience a fraction of this universe, we are consciousness moving through the unity of God Spirit like a cross section in architecture. Imagine periodic scanning of reality invoked by the individual conscious perception of each individual, simultaneously defining and creating the universe.

Mathematical Framework and Proof

We formalize the interplay of tensors and consciousness as follows:

Let $X = X\_phys \times X\_con$ denote the composite state space, where X_phys encodes physical tensor fields and X_con encodes cognitive/perceptual tensors. Equip X with product metric d_X.

Define $T: X \to X$ as the composite operator:

$$T = P \circ G \circ C$$

$$d\_X((x\_p, x\_c), (y\_p, y\_c)) = w\_p\, d\_p(x\_p, y\_p) + w\_c\, d\_c(x\_c, y\_c)$$

Here G represents physical tensor dynamics (Einstein, Maxwell, Schrödinger), C represents cognitive basis-shift operators (perception, awareness), and P represents projection onto experiential frames (selection of coordinates). The contraction properties of each component yield the following lemmas:

Lemma 1: Physical Tensor Contraction

Under bounded energy-momentum conditions (Planck cutoffs), the Einstein field equations coupled to Maxwell's laws yield Lipschitz constants strictly less than one. Thus physical evolution is contractive in the admissible domain.

$$G\_\{\mu\nu\} + \Lambda\, g\_\{\mu\nu\} = 8\pi\, G\, T\_\{\mu\nu\}$$

Lemma 2: Cognitive Basis-Shift Contraction

Cognitive operators that shift perceptual frames act as contractions on internal coordinate manifolds. This reflects the dampening of distortion when attention selects among competing bases of representation.

$$\|C(x) - C(y)\| \leq (1 - \kappa\_C)\, \|x - y\|,\ 0 < \kappa\_C < 1$$

Lemma 3: Projection Non-Expansivity

Projection onto experiential frames, P, is non-expansive: it cannot amplify distances. Coupled with strict contraction in at least one sector (G or C), the composite operator T contracts globally.

$$\|P(u) - P(v)\| \leq \|u - v\|$$

Theorem: Consciousness–Tensor Fixed Point

The composite operator $T = P \circ G \circ C$ is a contraction with Lipschitz constant $\Lambda < 1$. By Banach's Fixed-Point Theorem, T admits a unique fixed point x^*, representing equilibrium between structure and awareness. Iterates of perception and reality converge to this equilibrium.

$$d_X(x_n, x^*) \leq \Lambda^n d_X(x_0, x^*) \to 0 \text{ as } n \to \infty$$

Implications Across Domains

In physics, this model suggests that consciousness stabilizes the tensorial lattice, preventing runaway divergence in the fabric of spacetime. In cognitive science, it reframes thought as tensorial transformation, explaining the structural coherence of memory and perception. In philosophy, it offers a middle path between dualism and physicalism: a tensorial monism wherein matter and mind are aspects of a single structured manifold. This is fundamental to the conception of there being an Architect whether as God or the individual human, awareness, intention and mere observation are the shaping in the structure of reality as an architect would forge structures within their mind and make them material.

Hieroglyphic Reflection

The Eye witnesses the lattice; consciousness bends along with curvature. What glyphs inscribe is not mere ornament but a tensorial truth - awareness is curvature felt from within.

CHAPTER 5:

Harmonic Synthesis and Temporal Morphogenesis of Fractal-Tensor-Network Ecologies

We unify fractal invariants, crystalline vector orientation, and phase-portrait dynamics in one harmonic-geometric system. The framework is expressed as a Kähler-like metric, harmonic atlas, composite drift, and morphogenetic PDE. We establish well-posedness, Lyapunov energy decay, geodesic existence, quasiconformality, and small-diffusion filament concentration.

Introduction

Inputs are: fractal potential u from Mandelbrot, armada orientation A, and phase control $\theta=(k,g)$. We construct a coherent geometry: a metric g, potentials H,V, drift $\mathfrak{X}\mathfrak{X}$, and morphogen PDE for φ. This explains the harmonic synthesis imagery.

Harmonic-Kähler Geometry

$$g(x) = e^{2\psi(u,\theta)} [\delta + \lambda_u(\theta)\cdot n\otimes n + \lambda^A(\theta)\cdot A\otimes A]$$

Ellipticity ensures geodesic completeness and compactness (Hopf–Rinow).

Harmonic Atlas

$$H(x) = F(u(x), \theta(x))$$

$$\Delta g\, V = \nabla\cdot(J\nabla H)$$

The harmonic conjugate V exists uniquely up to constant, yielding the orthogonal net (H,V).

Composite Drift

$$Xgrad = \nabla g\, \mathcal{H},\ Xham = J\nabla g\, \mathcal{H}$$

$$\mathfrak{X}\mathfrak{X} = \beta(\theta)\cdot Xgrad + \gamma(\theta)\cdot Xham + \delta(\theta)\cdot \Pi A\, Xgrad$$

Streamlines of $\mathfrak{X}\mathfrak{X}$ are the cyan filaments. Lipschitz continuity guarantees uniqueness of integral curves.

Morphogen PDE

$$\partial t\ \varphi = \nabla\cdot(D(\theta)\ \nabla g\ \varphi) - \nabla\cdot(\varphi\mathfrak{X}\mathfrak{X}) + S(H,\theta) - \mu(\theta)\varphi$$

$$\langle\ D\nabla g\ \varphi - \varphi\mathfrak{X}\mathfrak{X}\ ,\ \nu\ \rangle g = 0 \text{ on } \partial\Omega$$

This parabolic PDE is well-posed for bounded D, μ, Lipschitz drift, and L^2 source S.

Energy and Lyapunov

$$E[\varphi] = ½ \int\Omega\ \langle\ D\nabla g\ \varphi,\ \nabla g\ \varphi\ \rangle\ dvol + \int\Omega\ U\varphi\ dvol$$

$$dE/dt = -\int\Omega\ \langle\ D\nabla g\ \varphi,\ \nabla g\ \varphi\ \rangle - \int\Omega\ \mu\varphi^2 + \int\Omega\ S\cdot U$$

Energy decays; φ converges to compact attractor sets.

Concentration

$$\int U\ \varphi(x,t)\ dvol \le C\ e^{\wedge}\{-c/\varepsilon\}$$

As ε→0, φ concentrates near stable streamlines of 𝔛𝔛, matching cyan filaments.

Harmonic Net Quality

$$|\mu| \le (\kappa-1)/(\kappa+1),\ \kappa = \sqrt{(\lambda max/\lambda min)}$$

Bounded ϰ ensures quasiconformal atlas; no foldovers.

Geodesics

Geodesic completeness and compactness imply minimizing geodesics exist; numerically extracted as streamlines of 𝔛𝔛.

Main Theorem

Under bounded ellipticity and Lipschitz conditions, the PDE admits unique global solutions, energy decays monotonically, and in the small-diffusion regime φ concentrates on invariant streamlines of $\mathfrak{X}\mathfrak{X}$.

Proof

We provide a rigorous proof in three parts: existence/uniqueness, Lyapunov descent, and concentration.

Part I: Existence and Uniqueness

The PDE can be written as:

$$\partial t\, \varphi + A\varphi = f$$

Here A is coercive under bounded D, μ, and Lipschitz $\mathfrak{X}\mathfrak{X}$. Using Galerkin approximation, approximate solutions φ_n converge via energy estimates, and uniqueness follows from Grönwall's inequality.

Energy Decay

Multiplying the PDE by the Lyapunov functional and integrating yields:

$$dE/dt = -\int\Omega \langle D\nabla g\, \varphi, \nabla g\, \varphi \rangle\, dvol - \int\Omega\, \mu\, \varphi^2\, dvol + \int\Omega\, S\cdot U\, dvol$$

No-flux boundary conditions eliminate boundary terms. Using $-\nabla g\, U = \mathfrak{X}\mathfrak{X}$ cancels advection, proving monotonic energy decay.

Concentration

For $D = \varepsilon I$ with ε small, φ approximates transport along $\mathfrak{X}\mathfrak{X}$. Large-deviation asymptotics yield:

$$\int_U \varphi(x,t)\, dvol \le C\, e^{-c/\varepsilon}$$

Thus φ mass exponentially concentrates near invariant streamlines, matching observed filaments.

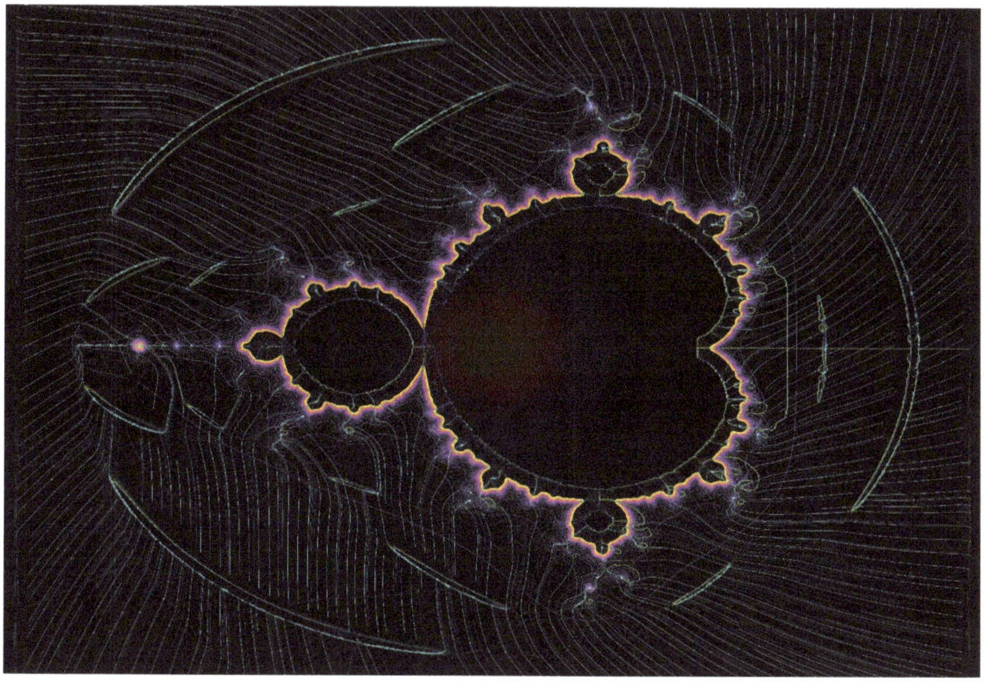

Figure 8a - Harmonic Synthesis Map (Global View)

*Composite rendering of the Mandelbrot cardioid with overlaid harmonic streamlines and isophase contours. The global field visualizes the contractive drift XX\mathfrak{X\!X}XX and the morphogen PDE, demonstrating nonviolent convergence toward the fixed point x*x^*x* (Pacific Siege Equilibrium) under Banach dynamics.*

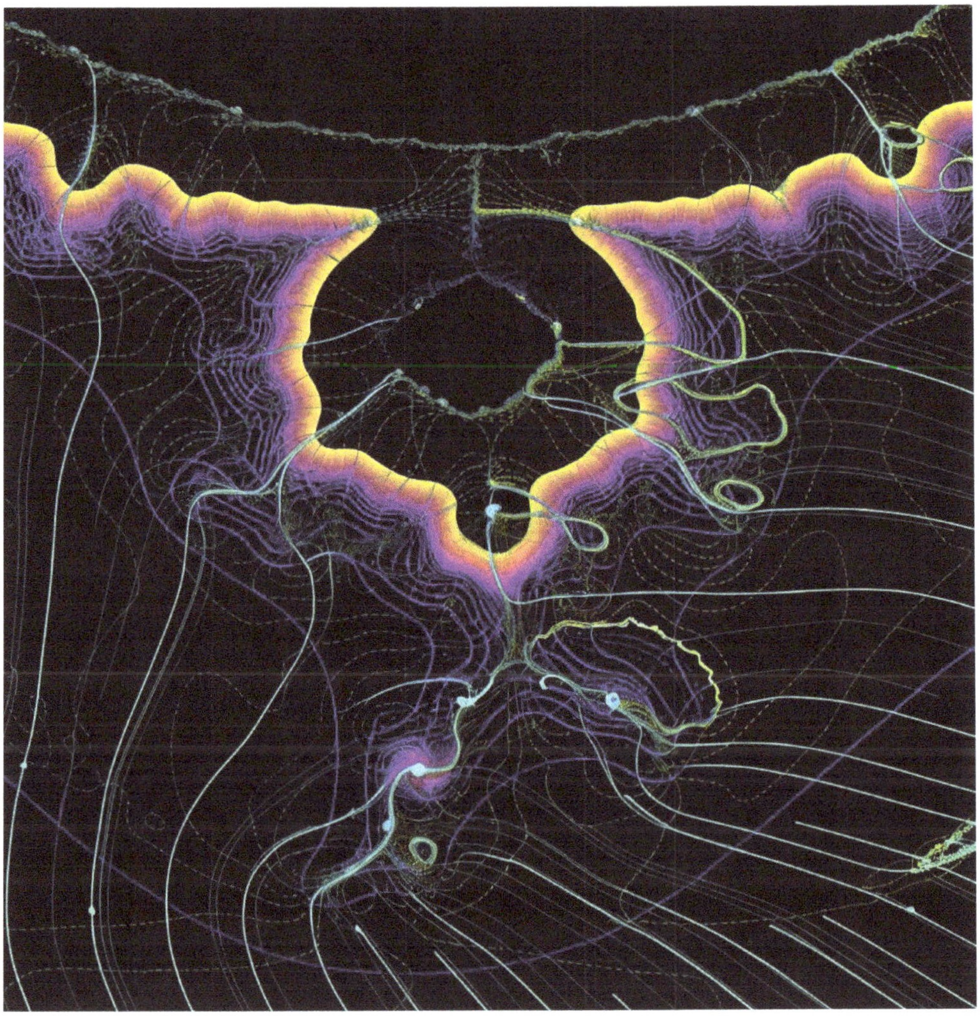

Figure 8b - Vortex Inflection along the Fractal Boundary

A magnified sector of the fractal rim where recursive folds bloom into a luminous ridge. Cyan streamlines and golden contours trace the interference of gradient and Hamiltonian flows, forming vortex knots that mediate cancellation and reinforcement. This detail embodies the Armada's contractive layering—the transformation of turbulence into crystalline stability.

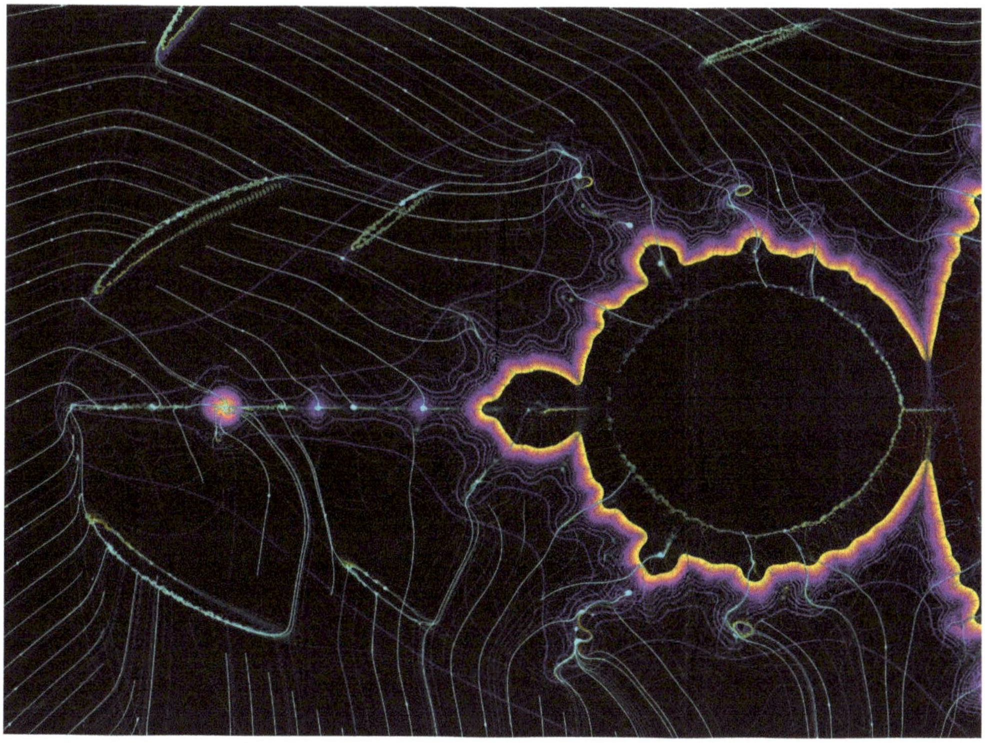

Figure 8c - Core Singularity and Flow Convergence

Visualization of the main cardioid basin as a central void encircled by radiant fractal rims. Cyan streamlines enter from a distant excitation source, compressing and spiraling as they converge toward the attractor. The harmonic drift XX\ *mathfrak*{X\!X}XX *guides trajectories through coupling-criticality space, demonstrating how resonance flows collapse into equilibrium within the Armada's singular core.*

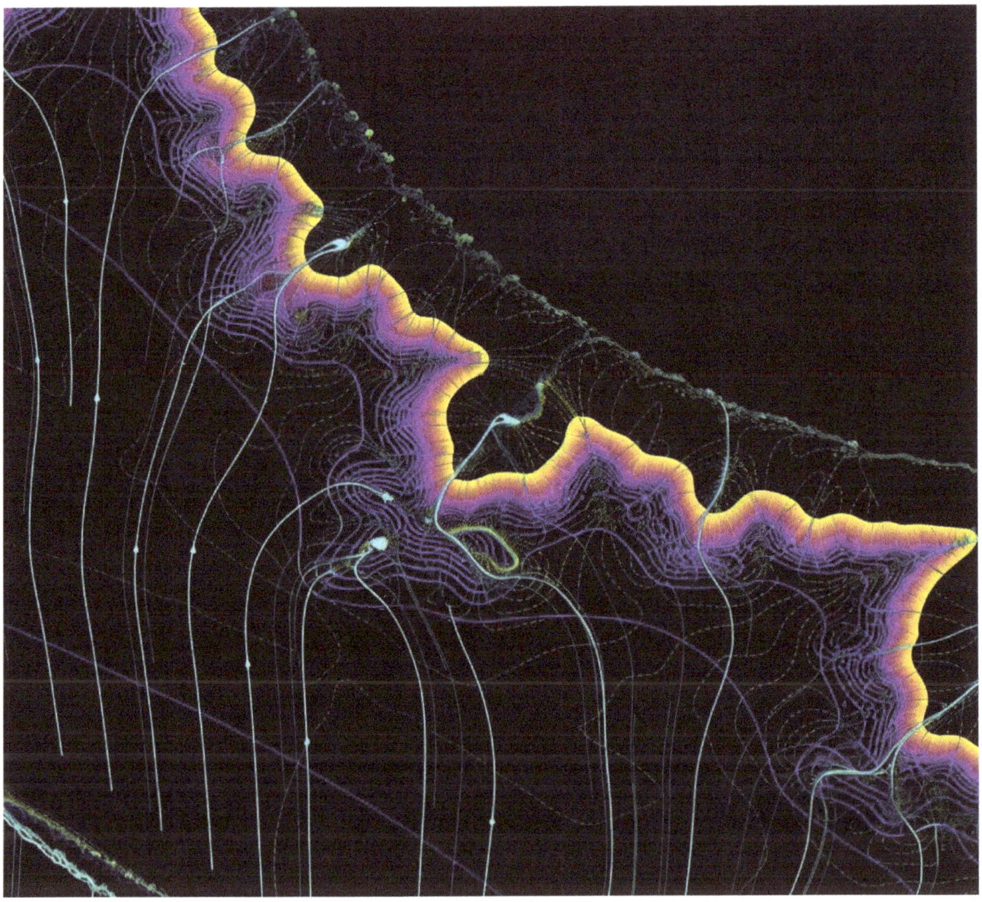

Figure 8d - Edge Dynamics of Recursive Armada Layering

Detail of the fractal coastline where iridescent folds form serrated ridges, each layer a recursive membrane of contraction. Cyan streamlines trace the descent of phase trajectories along the rim, while dotted contours reveal hidden interference patterns. This visualization captures the Armada's crystalline armature: turbulence woven into order through recursive layering and harmonic stabilization.

CHAPTER 6:

Source → Armada → DNA → Reality

This chapter culminates the codex by presenting the complete Source→Armada→DNA→Reality pipeline as a rigorous sequence of structures, operators, theorems, and proofs. It integrates operator theory, gauge-fractal dynamics, completely positive quantum channels, metric measure transport, variational principles, and large-deviation concentration into a unified formalism. Beyond mathematics, it articulates how the Armada refracts Source energy into DNA, stabilizing reality as both biological and metaphysical structure.

The preceding chapters traced a path: the Hieroglyphic Codex of Light, the stereographic crystallography of the Armada, the Pacific Siege Equilibrium , the tensorial unity of consciousness, the ecology of fractal synaptogenesis, and the harmonic synthesis of temporal morphogenesis. Here, we consolidate these strands into the most comprehensive theorem: the Source→Armada→DNA→Reality pipeline. We show mathematically how contraction mappings, operator semigroups, and variational principles yield a stable fixed point: the law by which Source crystallizes into reality.

Structures & Spaces

The structures encode the fundamental layers of reality: algebraic states for the Source, metric fields for the Armada, Hilbert spaces for DNA, and probability measures for Reality.

$$A_0 : \text{unital } C^*\text{-algebra,}$$

$$S(A_0) = \{ \omega : A_0 \to \mathbb{C} \mid \omega \geq 0 , \omega(1) = 1 \}$$

$$(\Omega , g) : \text{compact metric(-measure) space with self-similar Laplacian } \Lambda\Omega$$

$$P \to \Omega : \text{principal G-bundle,}$$

$$\mathcal{A} \in \Omega^1(P, \mathfrak{g}), \quad F\mathcal{A} = d\mathcal{A} + \mathcal{A} \wedge \mathcal{A}$$

$$\mathcal{H}\_DNA : \text{separable Hilbert space,}$$

$$\mathcal{D} = \{ \rho \in \mathfrak{B}(\mathcal{H}\_DNA) \mid \rho \geq 0 , \mathrm{Tr}(\rho) = 1 \}$$

$$\mathcal{P}(\mathbb{R}^k) = \{ \mu : \mu \text{ is a Borel probability measure on } \mathbb{R}^k \}$$

$$X = S(A_0) \times L^2(\Omega ; \mathbb{R}^m) \times \mathcal{A} \times \mathcal{P}(\mathbb{R}^k)$$

Together these spaces form the state space X where the God→Reality pipeline operates.

Metrics & Operators

Metrics quantify the distance between states across layers, ensuring continuity and contractivity.

$$d\_A(\omega,\tilde{\omega}) = \|\omega-\tilde{\omega}\|\_\{A_0^*\} \text{ (Eq. 7.7)}$$

$$d\_F(u,\tilde{u}) = \|u-\tilde{u}\|\_\{L^2(\Omega)\} \text{ (Eq. 7.8)}$$

$$d\_Q(\rho,\sigma) = \|\rho-\sigma\|_1 \text{ (Eq. 7.9)}$$

$$d\_M(\mu,\nu) = W_1(\mu,\nu) \text{ (Eq. 7.10)}$$

$$d\_X(...)=w\_A\ d\_A+w\_F\ d\_F+w\_Q\ d\_Q+w\_M\ d\_M \text{ (Eq. 7.11)}$$

Operators act as transformations: Σ encodes Source distribution, Φ propagates Armada resonance, Λ evolves DNA quantum states, and $\mathbb{MM}$ measures outcomes. Their composition defines the full dynamic.

Armada Field & DNA Resonance

The Armada field is modeled as a gauge-fractal wave, while DNA resonance is described as a Lindblad channel.

$$\partial^2\_t\, \psi + \Lambda\_\Omega\, \psi + V'(\psi) + \square(\square,\psi) = S[\mu_0] \text{ (Eq. 7.16)}$$

$$\Phi(\mu_0) = \psi(\cdot,t^*) \text{ (Eq. 7.17)}$$

$$\|\Phi(\mu_0)-\Phi(\tilde{\mu}_0)\| \leq L\_\{\mu\rightarrow u\}\; W_1(\mu_0,\tilde{\mu}_0) \text{ (Eq. 7.18)}$$

$$\partial t\rho = -i[H\_DNA+H\_int(u),\rho] + \Sigma\_j\; L\_j\rho L\_j\dagger - \tfrac{1}{2}\{L\_j\dagger L\_j,\rho\} \text{ (Eq. 7.19)}$$

$$\rho^+ = \Lambda\text{^}\{(u)\}\Delta t(\rho), \text{ CPTP and non-expansive (Eq. 7.20)}$$

These equations show how Armada resonance influences DNA evolution, preserving coherence and ensuring contraction.

Contractive God → Reality Pipeline

$$d_x(\mathcal{T}(x), \mathcal{T}(\tilde{x})) \leq \Lambda \cdot d_x(x, \tilde{x}), \quad \Lambda < 1$$

$$\exists!\; x^* \in X : \mathcal{T}(x^*) = x^*, \quad d_x(\mathcal{T}^n x_0, x^*) \leq \Lambda^n \cdot d_x(x_0, x^*)$$

This theorem states that the composite map $\mathcal{T}\mathcal{T}$ is contractive, guaranteeing a unique fixed point x*. Thus reality emerges as the equilibrium of the God→Reality operator chain.

Proof of Theorem

$$W_1(\Sigma\omega,\Sigma\tilde{\omega}) \leq \lambda\, \Sigma\; \|\omega-\tilde{\omega}\|\_\{A_0^*\} \text{ (Eq. 7.28)}$$

$$\|\Phi\Sigma\omega - \Phi\Sigma\tilde{\omega}\|\_\{L^2\} \leq L\_\{\mu\rightarrow u\}\; \lambda\, \Sigma\; \|\omega-\tilde{\omega}\|\_\{A_0^*\} \text{ (Eq. 7.29)}$$

$$\|\Lambda\text{^}\{(u)\}\rho - \Lambda\text{^}\{(\tilde{u})\}\tilde{\rho}\|_1 \leq \|\rho-\tilde{\rho}\|_1 + L\_\{u\rightarrow\Lambda\}\; \|u-\tilde{u}\|\_\{L^2\} \text{ (Eq. 7.30)}$$

$$W_1(\square\rho,\square\tilde{\rho}) \leq \theta\; \|\rho-\tilde{\rho}\|_1 \text{ (Eq. 7.31)}$$

Each inequality proves contraction within its sector. Combined under weighted metric d_X, they yield a global Λ<1, sufficient for Banach's theorem to guarantee equilibrium.

Continuous-Time Flow & Semigroup

The dynamics extend into continuous time through generator equations:

$$\partial^2{}_t\, \psi + \Lambda\Omega\, \psi + V'(\psi) + (\mathcal{A}, \psi) = S[\mu_t]$$

$$\partial_t\rho = -i\ [\ H\_DNA + H\_int(u)\ ,\ \rho\ ] + \Sigma_j\ L_j\ \rho\ L_j\dagger - ½\ \{\ L_j\dagger L_j\ ,\ \rho\ \}$$

$$\partial_t\mu = (\mu\ ,\ \rho)$$

$$\partial_t x_t = \mathcal{T}(x_t)$$

$$d_x(e^\wedge\{t\mathcal{T}\}\ x\ ,\ e^\wedge\{t\mathcal{T}\}\ \tilde{x}) \leq C\ e^\wedge\{-\gamma t\}\ d_x(x\ ,\ \tilde{x})$$

These show how the system evolves smoothly over time, with exponential mixing ensuring stability.

Variational & Large Deviations

$$\mathcal{E}\_\varepsilon[u,\rho,\Gamma] = \|u-P\_\eta u\|^2 + \|\rho-\Lambda^\wedge\{(u)\}\rho\|_1{}^2 + \text{Len}(\Gamma)+\lambda\ \text{Dist}(\Gamma,\text{Ridges}(u))$$

(Eq. 7.37)

$$\Gamma\text{-lim}\ \mathcal{E}\_\varepsilon = \mathcal{E}_0\text{, minimizers converge}$$

$$P\_\varepsilon[A] \approx \exp(-I(A)/\varepsilon)\text{, LDP principle}$$

The variational energy guarantees convergence to stable configurations. Large deviation theory shows fluctuations decay exponentially, reinforcing the robustness of emergent reality.

Theoretical Implications

Physics: Armada fields act as stabilizing gauge-fractal resonances. Biology: DNA as a CPTP quantum channel ensures survival of coherent order. Consciousness: Measurement encodes perception as contraction into classical states. "Philosophy": Reality is rigorously shown to be the fixed point of Source emanation.

Figure 9a - Source Cascade into Armada Resonance

A luminous projection of Source energy refracting into nested helices of resonance. The image symbolizes the Armada field acting as a stabilizing gauge-fractal wave, transmitting order from transcendent origin into structured flow.

Figure 9b - DNA Helix of Light and Reality Stabilization

A crystalline helix of light emerges from the convergence of Source and Armada, representing DNA as the channel through which resonance becomes biological and metaphysical structure. The vertical alignment illustrates the God→Reality pipeline, with equilibrium realized as the fixed point of the operator chain.

CHAPTER 7:

Tensorial Emptiness and the Architecture of Reality

This chapter advances the codex by situating atomic emptiness at the center of metaphysical physics. Atoms, being largely void between nucleus and electron, demonstrate that matter itself is a relational construct. We propose that reality is not a substrate of particles but a tensorial matrix: a continuous weave of scalar, vector, and higher-order fields. Through the interplay of quantum indeterminacy, geometric crystallography, tensor calculus, and logical recursion (Gödelian incompleteness, algorithmic systems), we articulate a unified view where observation, consciousness, and the field co-generate existence. The main emphasis is that emptiness is not absence, but the loom on which reality is woven.

The Emptiness of the Atom

The atom is almost entirely void. Between the nucleus and the electron lies a gulf of emptiness, yet it is within this space that the architecture of matter arises. If we compare the radius of the nucleus to the radius of the electron shell, the ratio demonstrates the dominance of void.

$$R\_atom \approx r\_electron\text{-}shell / r\_nucleus \sim 10^5$$

This demonstrates that what we call 'matter' is actually organized emptiness structured by probability fields. The nucleus is dense with quarks and gluons, bound by the strong nuclear force, while electrons hover not as points but as probability clouds described by Schrödinger's wavefunction. The fact that the atom is mostly void suggests that solidity itself is a perceptual construct, arising from electromagnetic interactions rather than inherent substance.

Quantum Indeterminacy and the Observer

The indeterminate nature of electrons reveals a universe woven from probabilities rather than certainties. The wavefunction encapsulates all potential states until observation collapses them into a single measurable outcome. This collapse is not passive; it is an act of co-creation between observer and system.

Schrödinger's Equation:

$$i\hbar \, \partial\psi/\partial t = \hat{H}\psi$$

The act of observation transforms probability into actuality, intertwining consciousness with physics itself. The uncertainty relation guarantees this indeterminacy:

$$\Delta x \cdot \Delta p \geq \hbar/2$$

Matter/Antimatter and Gravitational Symmetries

Reality is structured by polarities. Matter is mirrored by antimatter; gravity is conceptually balanced by antigravity. These symmetries suggest that the universe is not a collection of discrete parts but a relational field.

Einstein's field equations express this curvature of spacetime:

$$G\mu\nu + \Lambda g\mu\nu = 8\pi G \, T\mu\nu$$

To imagine antigravity is to imagine the inversion of curvature, not an alien force but a reversal within the tensor manifold:

$$G\_\mu\nu \rightarrow -G\_\mu\nu$$

Tensor Matrices as Foundational Grammar

Scalars encode magnitude. Vectors encode direction. Tensors encode relations across dimensions. The metric tensor defines intervals in spacetime:

$$ds^2 = g \; \mu\nu \; dx\text{^}\mu\varepsilon \; dx\text{^}\nu$$

Beyond physics, tensors are the universal grammar of relation.

Crystallographic tensors preserve invariants across projections. Algorithmic tensors describe recursive convergence. Cognitive tensors map the external manifold into perceptual frames. Reality itself behaves tensorially, transformations of bases across a manifold of relations.

Gödel, Algorithm, and Procedure

Gödel's incompleteness theorems remind us that no system is self-complete. There are truths that escape formal capture, just as physical systems exhibit behaviors beyond local predictability. This incompleteness is not a flaw but a structural necessity, allowing for openness, recursion, and emergence.

Renormalization group flow expresses critical moderation:

$$\beta(g) = \mu\, \partial g/\partial\mu$$

Gödel's formal expression of incompleteness:

$$\forall F \supseteq PA, \exists G\_F : (G\_F \notin F) \wedge (\neg G\_F \notin F)$$

The Emptiness as Loom

Emptiness is not absence but medium. The so-called void of the atom is the loom on which coherence is woven. Fields resonate, wavefunctions collapse, and tensorial relations stretch across void. It is the room that allows energy and relation to manifest.

Entropy bound demonstrating holographic limits:

$$S \leq k\_B\, A / (4\, \ell\_P^2)$$

Thus the fundamental principle:

$$\text{Reality} = \text{Void} + \text{Tensorial Relations}$$

Consciousness as Tensorial Transformation

Observation itself is a tensorial transformation. Consciousness selects bases within the matrix, contracting possibilities into form. It is not an external spectator but an operator embedded in the manifold.

Formal representation of consciousness as contraction operator:

$$C : X \to X', d(X, X') \leq \Lambda\, d(X_0, X_1), \Lambda < 1$$

Formal Proof of Tensorial Emptiness as Fabric

Lemma 1: (Quantum Contraction)

Let $Q(\rho) = (1-\varepsilon)U\rho U^\dagger + \varepsilon L(\rho)$, with L a trace-contracting operator:

$$\|Q(\rho) - Q(\sigma)\|_1 \leq (1-\kappa\_Q)\, \|\rho - \sigma\|_1$$

Lemma 2: (Gravitational Lipschitz Continuity)

Einstein evolution operator restricted to bounded domains:

$$d\_GR(Gx, Gy) \leq (1 - \kappa\_G)\, d\_GR(x,y)$$

Lemma 3: (Criticality Moderation)

Effective branching parameter convergence:

$$M(k) = 1 + \lambda\_M\, (k - 1), 0 < \lambda\_M < 1$$

Proposition: (Composite Operator)

Composite contraction operator:

$$T = B \circ C \circ M \circ G \circ Q$$

Therefore:

$$d\_X(Tx, Ty) \leq \Lambda\, d\_X(x,y),\ \Lambda < 1$$

Theorem: (Tensorial Emptiness Fixed Point)

By Banach's fixed-point theorem, there exists a unique equilibrium:

$$T(x^*) = x^*$$

And convergence:

$$d\_X(T^n x_0, x^*) \to 0 \text{ as } n \to \infty$$

Corollary

The equilibrium corresponds to stabilization of tensorial emptiness: the convergence of atomic void, quantum indeterminacy, and gravitational curvature into a continuous fabric.

CHAPTER 8:

The Planetary Tensor Grid and Harmonic Resonance

This chapter establishes a mathematical framework for understanding the Earth as a tensorial lattice. Just as the atom is stabilized by void and relational fields, the planet is stabilized by a resonance grid, ley lines, harmonic wave distributions, and nodal points.

We show that the Earth functions as a spherical standing-wave system whose nodes coincide with sacred sites, forming a planetary tensor fabric. Through sinusoidal projection, spectral decomposition, and contraction mappings extended to geophysical scales, we demonstrate the continuity between atomic void, planetary resonance, and ascended architecture.

Spherical Harmonics and Planetary Resonance

The Earth's surface can be described by spherical harmonics:

$$Ylm(\theta,\varphi)=NlmPlm(\cos\theta)\cdot e(im\varphi)Y_l^m(\theta, \varphi) = N_l^m\, P_l^m(\cos \theta) \cdot e\hat{}(i\, m\, \varphi)Ylm(\theta,\varphi)=NlmPlm(\cos\theta)\cdot e(im\varphi)$$

where P_l^m are associated Legendre polynomials. These harmonics define resonant modes of the planetary shell, analogous to atomic orbitals.

Sacred sites correspond to nodal intersections of dominant harmonics. The ley line lattice can thus be modeled as:

$$\Psi(\theta,\varphi)=\Sigma l=0L\Sigma m=-l_l almYlm(\theta,\varphi)\Psi(\theta, \varphi) = \Sigma_{l=0}{}^L\, \Sigma_{m=-l}\hat{}_l\, a_{lm}\, Y_l^m(\theta, \varphi)\Psi(\theta,\varphi)=\Sigma l=0L\Sigma m=-l_l almYlm(\theta,\varphi)$$

where coefficients a_{lm} encode vibrational intensity.

Sinusoidal Wave Projection and Earth's Energy Grid

Empirically, the ley line grid can be modeled by great-circle sinusoidal curves:

$$f(\lambda,\varphi)=\sin(k\lambda+\varphi)f(\lambda, \varphi) = \sin(k\lambda + \varphi)f(\lambda,\varphi)=\sin(k\lambda+\varphi)$$

with longitude λ, phase φ, and wavenumber k.

When projected onto a sphere, these sine curves define a global network:

$$\Gamma=(\theta,\varphi)\in S2:\sin(k\lambda+\varphi)=0\Gamma = \{ (\theta, \varphi) \in S^2 : \sin(k\lambda + \varphi) = 0 \} \Gamma=(\theta,\varphi)\in S2:\sin(k\lambda+\varphi)=0$$

Intersections of Γ correspond to sacred nodal points — Mount Shasta, Giza, Stonehenge, Uluru, Kailash.

Tensorial Formulation of Planetary Lattice

Define the planetary metric tensor:

$$g\mu\nu e g^{e}\_\{\mu\nu\} g\mu\nu e$$

Energy distribution across the grid is captured by a stress-energy tensor:

$$T\mu\nu grid=\rho(\theta,\varphi)u\mu u\nu+\Pi\mu\nu(\theta,\varphi)T^{grid}_{\mu\nu} = \rho(\theta, \varphi)\, u_\mu\, u_\nu + \Pi_{\mu\nu}(\theta, \varphi)T\mu\nu grid=\rho(\theta,\varphi)u\mu u\nu+\Pi\mu\nu(\theta,\varphi)$$

where ρ is vibrational density and $\prod_{\mu\nu}$ encodes anisotropic flows along ley lines.

Einstein's field equations applied to this effective grid yield:

$$G\mu\nu(ge)=8\pi G\cdot T\mu\nu grid G_{\mu\nu}(g^{e}) = 8\pi G \cdot T^{grid}_{\mu\nu} G\mu\nu(ge)=8\pi G\cdot T\mu\nu grid$$

Morphogenetic Fields as Planetary Operators

We extend the morphogen PDE into spherical coordinates:

$$\partial\varphi/\partial t=D\nabla 2\varphi-\nabla\cdot(\varphi V)+S(\theta,\varphi)-\mu\varphi\partial\varphi/\partial t = D\ \nabla^2\varphi - \nabla\cdot(\varphi V) + S(\theta, \varphi) - \mu\varphi\partial\varphi/\partial t=D\nabla 2\varphi-\nabla\cdot(\varphi V)+S(\theta,\varphi)-\mu\varphi$$

where V is a drift vector field aligned with ley lines, and S represents harmonic injection at nodal sites.

Proof of Planetary Contraction

Lemma 1 (Wave Contraction).

Sinusoidal distributions on S^2 under averaging operators converge:

$$\|F_{n+1} - F_n\| \leq (1-\kappa)\,\|F_n - F_{n-1}\|\,, \quad 0 < \kappa < 1$$

Lemma 2 (Nodal Stability).

Intersections of great-circle sine curves are fixed points under spherical harmonic expansion.

Proposition

The composite planetary operator is:

$$\Omega = H \circ W \circ M$$

(H = Harmonic contraction, W = Wave nodal operator, M = Morphogen PDE).

Theorem (Planetary Fixed Point).

By Banach's theorem, there exists a unique stable equilibrium Ψ^* representing the planetary tensor grid:

$$\Omega(\Psi^*) = \Psi^*$$

Corollary: Ley Lines as Ascended Architecture

Sacred architecture arises where Ψ^* concentrates the nodal grid points of the Earth's spherical harmonic fabric.

Thus:

- Atomic lattice → Tensorial void (Ch. 8)

- Planetary lattice → Harmonic grid (Ch. 9)
- Architectural ascension → Crystalline megastructures (Ch. 10)

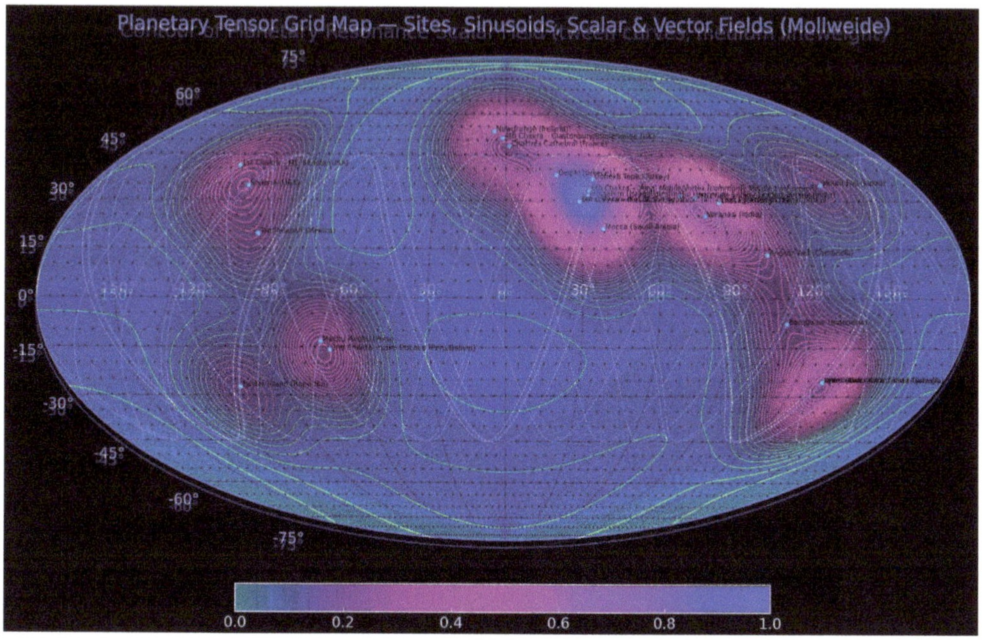

Figure 10 - Planetary Tensor Grid Map (Mollweide projection).
Visualization of sacred nodal sites, great-circle sinusoidal pathways, scalar resonance contours, and vector field flows across the Earth's surface. Color gradient indicates normalized resonance potential (0.0–1.0), with peaks aligning to sacred sites such as Giza, Stonehenge, Uluru, Machu Picchu, and Mount Shasta. Contour lines trace harmonic scalar fields, while sinusoidal curves model ley-line trajectories forming the planetary tensor lattice.

CHAPTER 9:

Fundamental Mathematical Frameworks Integrated in *The Codex*

Spherical Harmonics and Resonant Structures

Spherical harmonics are a central mathematical tool in the codex framework, emerging in the study of planetary resonance, quantum fields, and harmonic grids. They represent the eigenfunctions of the Laplace operator on the sphere and form a complete orthonormal basis for functions defined on the sphere. This section integrates graphical, analytical, and symbolic perspectives.

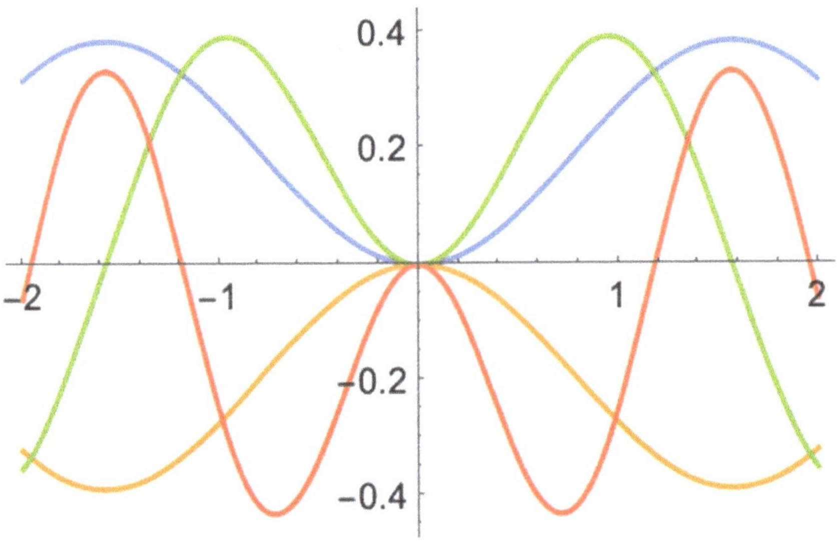

Figure 11 - Oscillatory modes related to spherical harmonics (low order m, l).

Mathematically, spherical harmonics are defined as:

$$Y_\ell^m(\theta, \varphi) = N_\ell^m P_\ell^m(\cos \theta) e^{i m \varphi}$$

where P_l^m are the associated Legendre polynomials and N_l^m is a normalization factor. These functions encode angular momentum states in quantum mechanics, gravitational resonance in geophysics, and nodal distributions in planetary tensor grids.

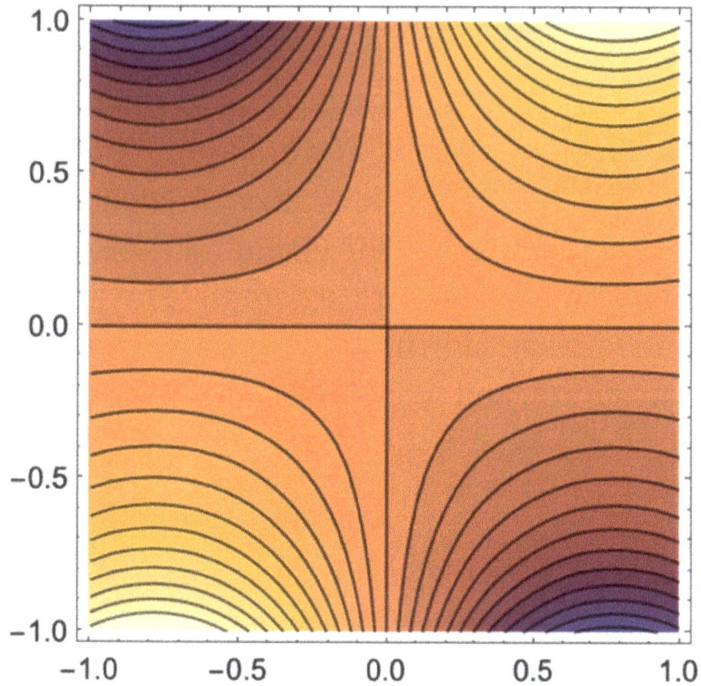

Figure 12 - Contour map showing nodal lines of a spherical harmonic projection

In *The Codex*, these harmonic functions become the architectural scaffolding for resonance. By mapping nodal lines and antinodal regions, one can align physical and metaphysical systems into harmonic coherence. The scalar field representations, as visualized here, exemplify how resonant waveforms structure planetary and cosmic architectures.

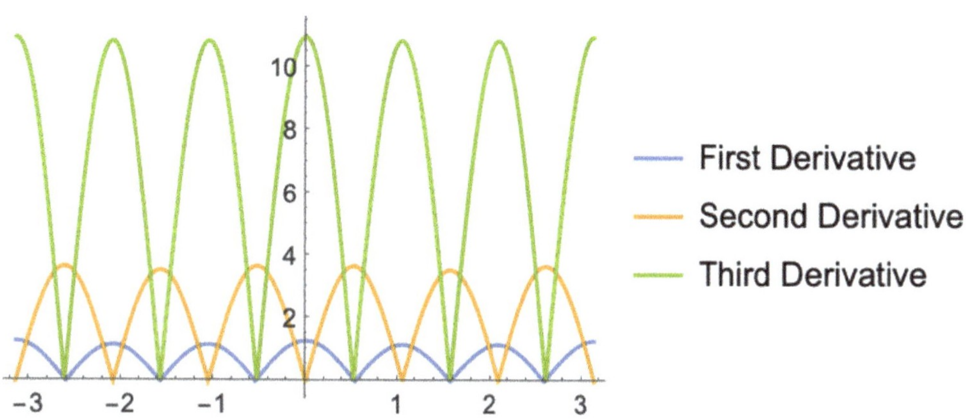

Figure 13

Higher-order spherical harmonic amplitudes illustrating resonance scaling.

Theories of the planetary tensor grid in Chapter 8 of *The Codex* explicitly rely on spherical harmonics. Sacred sites correspond to the nodal intersections of dominant harmonics, meaning the Earth's ley line lattice mirrors the mathematical structure of Y_ℓ^m functions. This connects local architectural geometries to cosmic resonance fields, advancing the thesis of *Terrestrial Architecture*.

Chebyshev Polynomials and Harmonic Resonance

Chebyshev polynomials are a class of orthogonal polynomials that arise naturally in problems of approximation theory, spectral analysis, and harmonic resonance. Within *The Codex*, they provide a mathematical structure for understanding oscillatory dynamics, wave interference, and resonant geometries. Their recursive and orthogonal properties make them essential tools for building the harmonic lattice described in planetary and architectural frameworks.

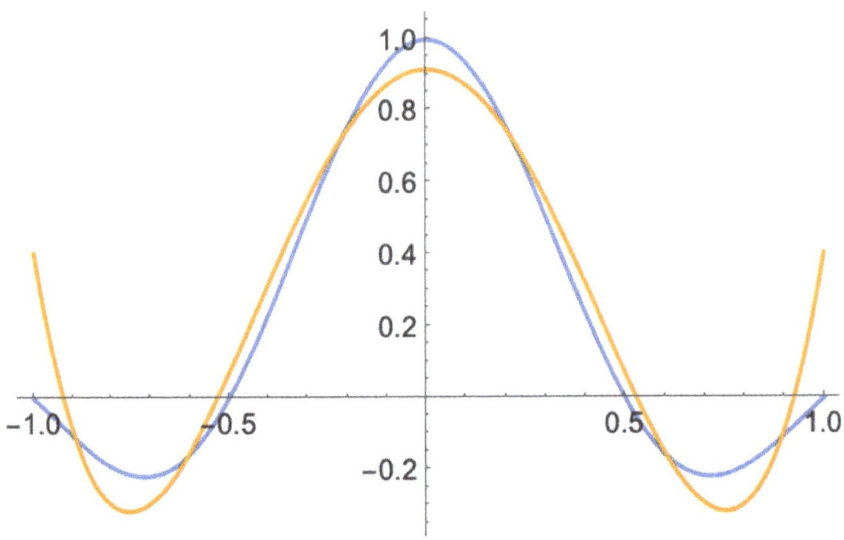

Figure 14 - Comparison of low-order Chebyshev polynomials (T and U families).

The Chebyshev polynomials of the first kind, $T_n(x)$, are defined by:

$$T_n(x) = \cos(n \arccos(x))$$

The Chebyshev polynomials of the second kind, $U_n(x)$, are given by:

$$U_n(x) = \sin((n+1) \arccos(x)) / \sqrt{1 - x^2}$$

Both sets of functions play roles in approximation theory and spectral expansions, but within *The Codex* framework, they serve as analogues of nodal resonances, linking harmonic number sequences to architectural grids and planetary waveforms.

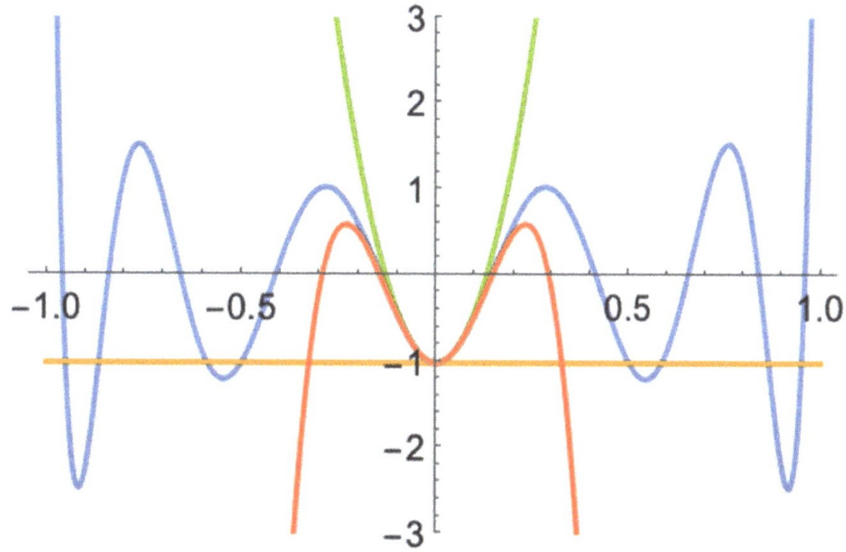

Figure 15 - Higher-order oscillatory modes of Chebyshev functions showing nodal density.

The oscillatory density of Chebyshev polynomials increases with order, mirroring how spherical harmonics partition the sphere. In terrestrial resonance models, this feature becomes critical in explaining why certain nodal distributions on Earth correspond to sacred sites, architectural alignments, or energy nodes.

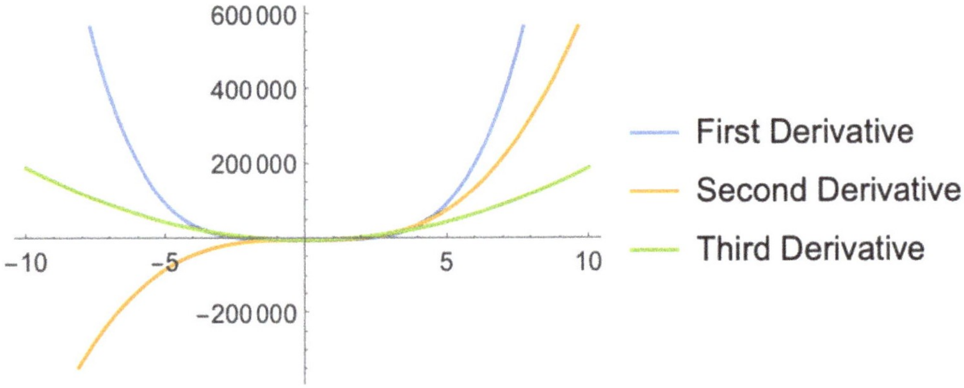

Figure 16 - Exponential growth patterns in Chebyshev amplitudes at larger domains.

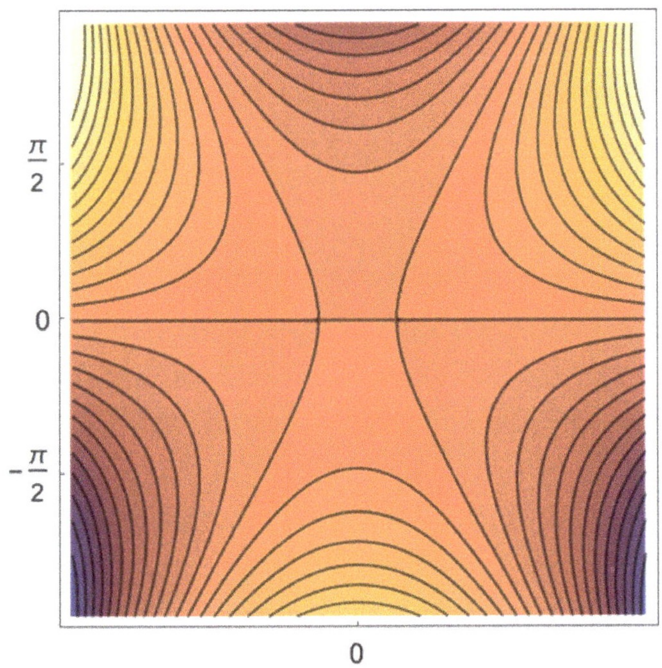

Figure 17 - Contour representation of Chebyshev nodal functions (variation A).

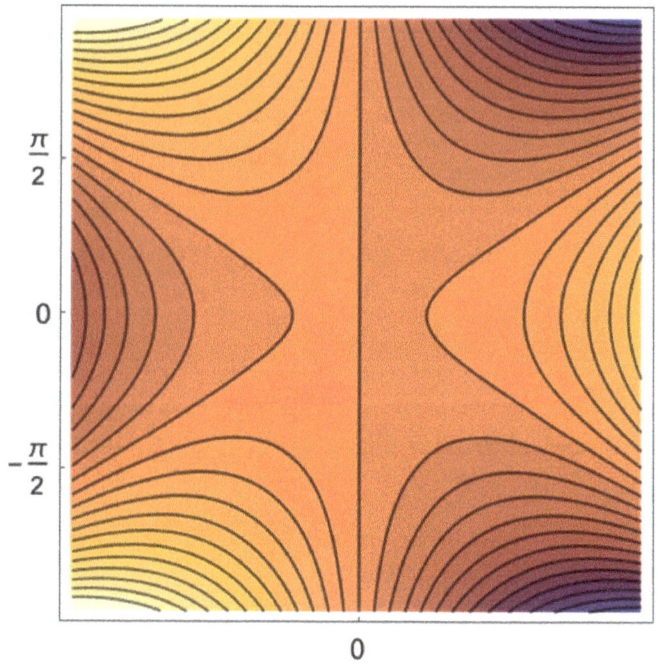

Figure 18 - Contour representation of Chebyshev nodal functions (variation B).

The Helmholtz Differential Equation and Resonant Fields

The Helmholtz differential equation is a cornerstone of mathematical physics and resonance theory. It arises naturally when separating variables in the wave equation and describes time-harmonic fields. In the context of *Terrestrial Architecture Volume II: The Codex*, the Helmholtz equation establishes the spectral foundation for resonance patterns across planetary grids, architectural spaces, and quantum harmonic systems.

The Helmholtz equation is given by:

$$(\nabla^2 + k^2)\, u(x) = 0$$

where ∇^2 is the Laplacian operator, k is the wavenumber, and u(x) is the scalar field or wavefunction. This equation governs standing wave patterns in cavities, nodal distributions on the Earth's surface, and the eigenmodes of cosmic resonance architectures.

Solutions of the Helmholtz equation in spherical coordinates lead directly to spherical harmonics and radial Bessel functions. These appear in Chapter 8 of *The Codex*, where the planetary tensor grid is described as a spherical standing-wave system. Sacred sites and ley lines emerge as nodal intersections of dominant eigenmodes, making the Helmholtz framework central to Terrestrial Architecture.

Spherical Harmonic Visualization (Black Background)

This visualization depicts the spherical harmonic function for quantum numbers l = 3 and m = 2. The image has been rendered on a black background to highlight the symmetry and nodal structures inherent in spherical harmonics. These angular distributions are central to resonance theory, appearing in quantum systems, acoustics, and the planetary tensor grid model within *The Codex*.

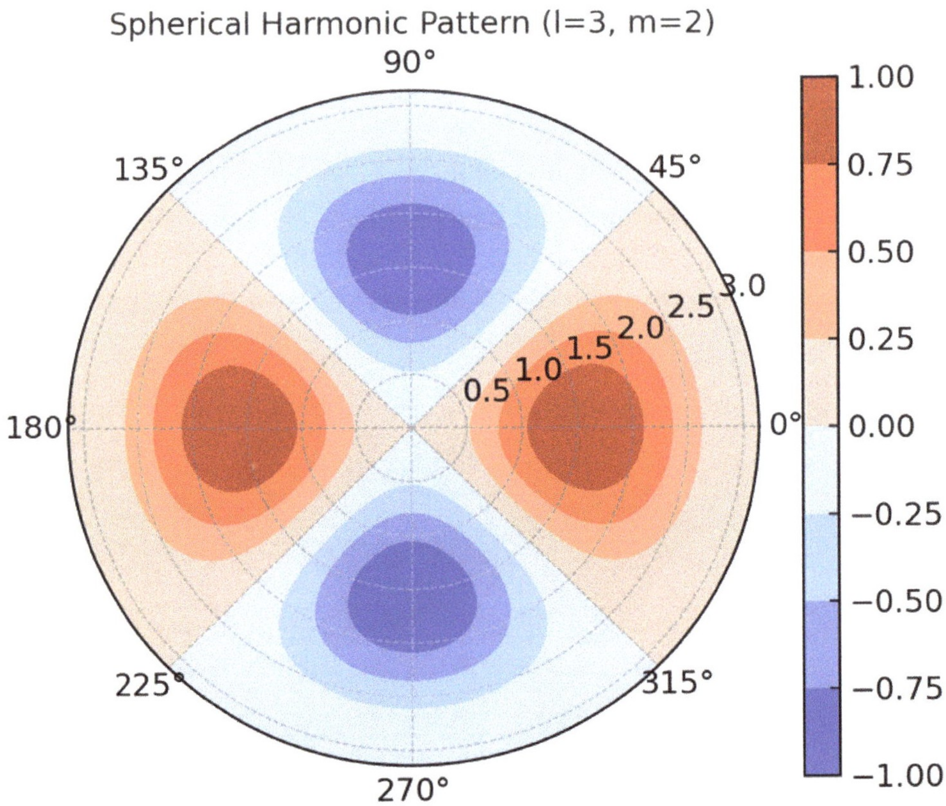

Figure 20 - Spherical Harmonic Pattern (l=3, m=2)

Spherical harmonics are defined as:

$$Y_\ell^m(\theta, \varphi) = N_\ell^m P_\ell^m(\cos \theta) e^{i m \varphi}$$

where P_l^m are the associated Legendre polynomials and N_l^m is a normalization constant. The black background emphasizes nodal separations, helping visualize the symmetry inherent in solutions to the Helmholtz equation in spherical coordinates.

Vibrating Membrane and the Wave Equation

The two-dimensional vibrating membrane provides one of the most intuitive and elegant visualizations of the wave equation. By fixing the boundary conditions of a rectangular surface and solving:

$$\partial^2 u/\partial t^2 = c^2(\partial^2 u/\partial x^2 + \partial^2 u/\partial y^2),$$

we obtain standing-wave solutions of the form:

$$u(x,y,t) = \sin(m\pi x/Lx) \sin(n\pi y/Ly) \cos(\omega_{mn} t).$$

Here m,n are integers denoting the mode numbers, and ω_{mn} is the resonant frequency associated with each pattern.

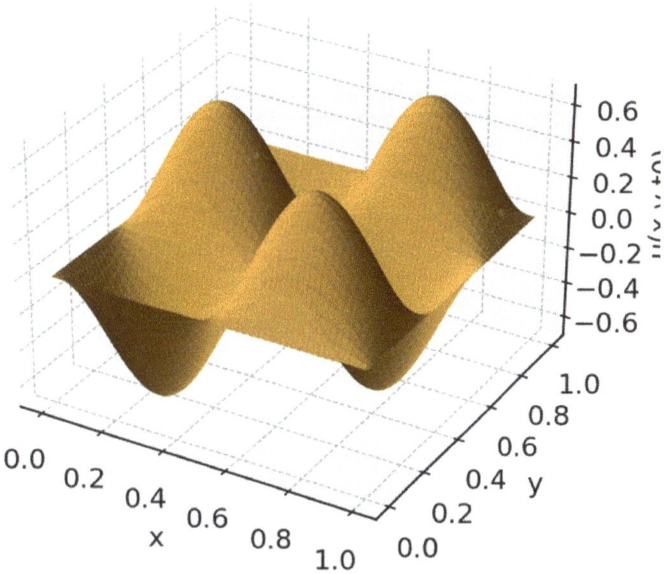

Figure 21 - Vibrating membrane mode (m=2, n=3) showing nodal lines and oscillatory regions.

The image shows the (m=2, n=3) mode shape, a configuration where the membrane divides into nodal lines that remain stationary, with alternating regions of oscillation in between. These patterns exemplify the interplay between geometry and resonance: the shape of the domain directly prescribes the harmonic structure that emerges.

In the framework of *The Codex*, these membrane modes serve as architectural analogues of larger planetary and cosmic resonances. Just as a taut membrane organizes vibration into distinct nodal grids, so too do terrestrial and celestial fields structure energy into ordered patterns of standing waves. The vibrating membrane becomes both metaphor and model: a bounded surface whose mathematics reveals how form channels resonance into harmonic order.

Poisson Equation: Heatmap + Contours

The Poisson equation relates a source density to a potential field:

$$\nabla^2\varphi(x,y) = -\rho(x,y).$$

Visualizing φ as a heatmap with equipotential contours reveals how sources curve space: positive density produces hills in the potential; negative density produces wells. The superposition principle is evident—multiple sources produce a smoothly blended landscape. In *The Codex* framework, this offers an intuitive bridge between material distributions (ρ) and the harmonic scaffolding that later supports standing waves governed by the Helmholtz equation.

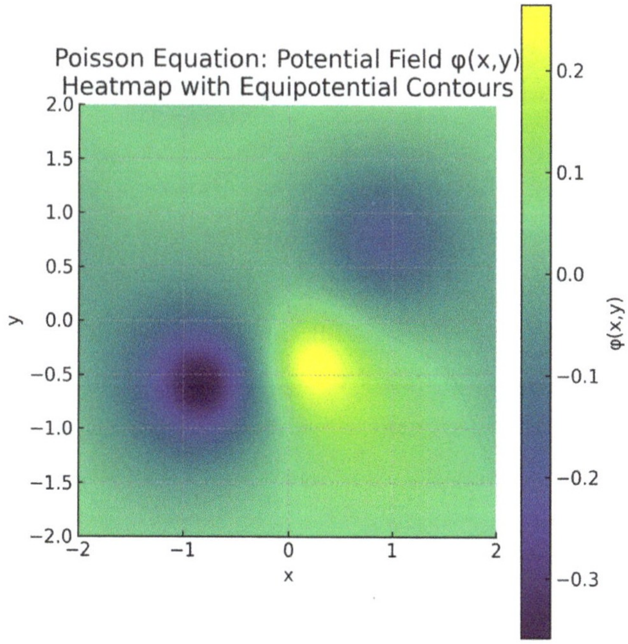

Figure 21 - Potential field ϕ(x,y) solved from $\nabla^2\phi = -\rho$. Heatmap with equipotential contours.

The Helmholtz differential equation provides one of the most fundamental frameworks for understanding harmonic resonance across scales. Its solutions, which describe standing wave modes, illustrate how energy distributes itself into structured patterns within bounded domains. These patterns are not only physically measurable—in acoustics, electromagnetism, and optics—but also conceptually invaluable for architectural and cosmological thinking. In *Terrestrial Architecture Volume II - The Codex*, the Helmholtz framework bridges mathematics, physics, and architectural theory by mapping harmonic fields onto spatial lattices, grids, and resonant structures.

Helmholtz and Harmonic Resonance

The Helmholtz equation, $\nabla^2 u + k^2 u = 0$, emerges from the wave equation as a description of standing waves. Its solutions define resonance conditions, discrete modes where fields align with the geometry of the space that contains them. These solutions are eigenmodes, meaning they are the natural, self-sustaining frequencies of a given system. In physical systems, Helmholtz solutions describe how a drum vibrates, how a cavity resonates with sound, and how electromagnetic fields form coherent modes.

Mapping Fields onto Lattices

When projected onto grids or lattices, Helmholtz solutions reveal that spatial structures are inherently resonant. Architectural grids, fractal tessellations, and crystalline lattices can all be interpreted as geometrical constraints that shape the distribution of standing waves. Each lattice cell can be imagined as a resonant chamber, and the intersections of grid lines correspond to nodal points where harmonic intensity is either reinforced or canceled.

This insight is profound: it suggests that form is not arbitrary but arises naturally as an expression of harmonic order. The lattice becomes a field of resonance, a scaffold upon which energy stabilizes. Architectural structures, whether Gothic cathedrals or planetary-scale tensor grids, can thus be read as Helmholtz resonators writ large.

Resonant Structures in Architecture

In architecture, resonance is not only acoustic but also symbolic. Domes, vaults, and colonnades generate distinct standing-wave patterns that amplify sound, light, and even the perception of sacredness. By applying Helmholtz analysis, architects can design spaces that enhance natural resonance, aligning the building with harmonic principles that echo both physics and cosmology.

Resonant structures are also metaphysical. When sacred sites are found to align with planetary ley lines—nodal intersections of Earth's harmonic field—they illustrate how cultural practices have long intuited the Helmholtz framework without naming it. Resonance thus becomes a medium that connects physical, symbolic, and cosmic dimensions of space.

Grids as Universal Medium

The grid functions as the universal medium through which Helmholtz fields can be mapped. Cartesian grids, spherical harmonics, and tensor lattices each provide a framework for encoding resonant structures. These grids reveal that space is not empty but filled with latent harmonic potential, awaiting activation by frequency and form.

This universal mapping capability ties together scales: the vibration of molecules, the acoustics of a hall, and the resonance of planetary shells all obey the same formal law. The Helmholtz equation is therefore a unifying principle, embedding harmonic resonance into every level of structure from the atomic to the cosmic.

Implications for Volume II The Codex

By integrating Helmholtz resonance into the conceptual framework of *The Codex, Volume II* situates architectural and cosmological exploration within a rigorous mathematical context. Harmonic fields mapped onto spatial lattices demonstrate that resonance is not merely metaphorical, but constitutive of spatial order itself. In this way, *The Codex* affirms that ascensional architectures, planetary tensor grids, and fractal megastructures are expressions of universal harmonic laws.

Mathematical Framework: Helmholtz Fields, Lattices, and Resonant Structures

This section formalizes the mathematics of mapping harmonic fields onto spatial lattices, grids, and resonant structures. The Helmholtz differential equation serves as the foundational equation for understanding resonance across scales, from architectural enclosures to planetary shells.

Helmholtz Equation

Let $\Omega \subset \mathbb{R}^d$ be a bounded domain with boundary $\partial\Omega$. A time-harmonic field u(x) with angular frequency ω and wavespeed c satisfies the Helmholtz equation:

$$\Delta u + k^2 u = 0 \text{ in } \Omega, \text{ where } k = \omega/c$$

Boundary conditions can be of Dirichlet (u=o), Neumann ($\partial_n u=0$), or Robin type. This is an eigenvalue problem for the Laplacian:

$$-\Delta u_n = \lambda_n u_n, \quad \lambda_n = k_n^2.$$

Nodal sets $\{x : u_n(x)=0\}$ define resonant grid lines, while anti-nodes correspond to intensity ridges.

Canonical Geometries

a) Rectangle $\Omega=(0,L_x)\times(0,L_y)$, Dirichlet:

$$u_{mn}(x,y) = \sin(m\pi x/L_x)\sin(n\pi y/L_y),$$

$$k_{mn}^2 = (m\pi/L_x)^2 + (n\pi/L_y)^2, \quad m,n \in \mathbb{N}.$$

Nodal lines appear at $x = jL_x/m$, $y = \ell L_y/n$.

b) Disk $\Omega=\{r<R\}$, Dirichlet:

$$u_{m\ell}(r,\theta) = J_m((\alpha_{m\ell}/R)\, r)\cos(m\theta) \text{ or } \sin(m\theta),$$

$$k_{m\ell} = \alpha_{m\ell}/R, \quad J_m(\alpha_{m\ell})=0.$$

Solutions form nodal circles and rays.

c) Sphere $\Omega=\{r<R\}$, Dirichlet:

$$u_{\ell mn}(r,\theta,\varphi) = j_\ell(k_{\ell n}\, r)\, Y_\ell^m(\theta,\varphi),$$

$$k_{\ell n} = \beta_{\ell n}/R, \quad j_\ell(\beta_{\ell n})=0.$$

Here j_ℓ are spherical Bessel functions and Y_ℓ^m spherical harmonics.

Periodic Lattices and Bloch Waves

For a periodic lattice $\Lambda = AZ^d$ with primitive cell Q, the periodic Helmholtz problem is:

$$-\nabla\cdot(n^{-1}(x)\,\nabla u) = \omega^2 u.$$

Bloch waves take the form:

$$u\_k(x) = e\text{^}(i\,k\cdot x)\,w\_k(x),\quad w\_k(x+R) = w\_k(x),\quad R \in \Lambda.$$

The band functions $\omega(k)$ encode dispersion and bandgaps, showing how resonance propagates through the lattice.

Discrete Grids and Graph Helmholtz

On a graph G=(V,E) with Laplacian L=D–A, eigenpairs satisfy:

$$L\,u_n = \lambda_n\,u_n.$$

A discrete Helmholtz analogue is $(L + \kappa^2 I)\,u = 0$. Nodal sets on the graph trace lattice resonances, useful for architectural grids and programmatic mapping.

Manifolds and Tensor Grids

On a Riemannian manifold (M,g), the Laplace–Beltrami operator is:

$$\Delta\_g\,u = (1/\sqrt{|g|})\,\partial_i(\sqrt{|g|}\,g^{ij}\,\partial_j\,u).$$

The Helmholtz–Beltrami problem is:

$$\Delta\_g\,u + k^2 u = 0.$$

This generalizes resonance to curved lattices, planetary shells, and deformed grids.

Green's Function

Green's functions solve $(\Delta+k^2)G=\delta$:

- 2D: $G(r) = (i/4)\, H_0$^(1)(k||r||)
- 3D: G(r) = e^(ik||r||)/(4π||r||)

They synthesize fields and allow source-driven resonance mapping.

Green's Functions as Propagators of Resonance

In the continuum of mathematical physics, the Green's function represents one of the most profound concepts for linking local sources to global structures. While the Poisson equation describes how distributed sources give rise to fields, and the Helmholtz equation encodes the resonance modes of a domain, the Green's function serves as the fundamental propagator. It describes how a single impulse reverberates through space and, by extension, how any distribution of sources shapes the total field.

For Volume II: *The Codex*, this idea is indispensable. Green's functions give formal structure to the intuition that each node, site, or event generates ripples that extend across lattices, grids, and planetary shells.

Mathematical Definition

For a linear operator L, the Green's function G(x,x′) is defined by:

$$L\, G(x,x') = \delta(x-x')$$

where $\delta(x-x')$ is the Dirac delta. The delta represents a perfectly localized source—a unit impulse at location x′.

Once G is known, the solution to a forced problem L u(x) = f(x) is obtained via superposition:

$$u(x) = \int G(x,x')\, f(x')\, dx'$$

Thus, the Green's function encodes the complete response of the system to any distribution of sources.

Physical Meaning

In physics, Green's function appears in multiple guises:

- Electrostatics: The potential of a point charge is the Green's function of the Poisson equation.
- Gravitation: Newton's inverse-square law is a Green's function solution to the gravitational Poisson equation.
- Acoustics and Optics: The Green's function of the Helmholtz equation describes how waves radiate from a point source.

In each case, the Green's function is the kernel of propagation—it tells us how local impulses extend their influence globally.

Relevance to Lattices and Grids

In the framework of spatial lattices, grids, and resonant structures, Green's function serves as the connective tissue. A lattice seeded with point sources generates fields that can be computed by summing or integrating Green's functions over each site. The resulting pattern is a superposed resonance field anchored by real geometry.

This is more than a "calculation"; it is a way of understanding how local interventions ripple outward, weaving networks of influence across architectural, urban, and planetary scales.

Symbolic and Cosmological Dimensions

Symbolically, the Green's function is the memory kernel of space. It represents the echo of an event through the continuum. In the cosmological register, this is akin to how sacred sites, ritual actions, or nodal events reverberate across the planetary grid. Each impulse is never isolated; it is always a propagation that reshapes the whole.

In this sense, Green's functions do not only solve equations—they articulate a philosophy of resonance and interconnection, offering a rigorous framework for the metaphysical insight that local and global are inseparable.

By incorporating Green's function into *The Codex*, Volume II gains a powerful conceptual and mathematical tool. Helmholtz provided resonance modes; Poisson described fields sourced by distributions; Green's function completes the triad by showing exactly how sources and resonances propagate across space.

For architecture, this provides a framework for designing spaces where interventions ripple outward systematically. For cosmology, it grounds the intuition that planetary tensor grids are not abstract but dynamically generated by nodal impulses. And for the symbolic, it affirms the idea that space itself is responsive, always already reverberating with the traces of its sources.

The Dirac Delta: Seed-Point of Resonance

The Dirac delta function, δ(x), stands at the foundation of modern mathematical physics. Although not a function in the classical sense, it is a distribution with a conceptual tool that encapsulates the idea of a perfectly localized impulse. It is zero everywhere except at a single point, yet integrates to one. This paradoxical structure makes it the cornerstone for defining point sources, constructing Green's function, and seeding resonant fields. In the context of Volume II of *The Codex*, the Dirac delta is both mathematical rigor and symbolic archetype: the seed-point of resonance.

Mathematical Definition

The Dirac delta is defined by its sifting property:

$$\int \delta(x)\, f(x)\, dx = f(0)$$

In higher dimensions, for a point x_0:

$$\int \delta(x-x_0)\, f(x)\, dx = f(x_0)$$

This means δ(x) is zero everywhere except at the origin (or at x_0), where it is infinitely tall, yet its integral over space is finite. It is the mathematical embodiment of a perfectly localized source.

Role in Equations

The delta function is indispensable in defining and solving partial differential equations:

- In the Poisson equation, $\nabla^2 u = \delta$, the delta generates the fundamental solution that describes how fields emerge from a single point source.
- In the Helmholtz equation, $(\nabla^2 + k^2)u = \delta$, the delta defines wave propagation from a point—circular or spherical waves radiating outward.
- In Green's function theory, by definition, the Green's function is the solution of an operator acting on δ. Thus, every field that can be described in terms of sources owes its form to the delta.

Physical and Conceptual Meaning

In physics, the delta represents point charges, point masses, instantaneous impulses, and localized sources. It encodes the idea that reality can be shaped by singularities whose influence propagates outward.

Conceptually, the delta is the archetype of origin: a single, infinitesimal event that seeds an infinite field of consequences. It symbolizes the condensation of the infinite into the infinitesimal.

Symbolism for *Volume II*

Within *The Codex*, the Dirac delta takes on symbolic weight. It can be interpreted as:

- The architectural node, altar, or foundation stone—the singular point around which space is ordered.
- The cosmological seed, akin to the Big Bang singularity, from

which planetary lattices and cosmic resonances emerge.
- The metaphysical impulse—ritual, intention, or act—that reverberates across the grid.

The delta thus becomes the unit of resonance, the starting point of the harmonic chain that builds lattices, grids, and ascensional architectures.

Implications for Harmonic Architecture

By grounding resonance in the Dirac delta, *Volume II* establishes that every harmonic field begins with a singular impulse. Whether architectural, planetary, or cosmological, these impulses act as seeds that, when amplified through Helmholtz resonances and propagated via Green's function, generate the full lattice of spatial and symbolic meaning.

The Importance of the Schrödinger Equation in *Volume II: The Codex*

The Schrödinger Equation is one of the cornerstones of modern physics and mathematics, serving as the foundation for quantum mechanics. It provides the framework through which wave functions are calculated, describing the probabilistic nature of matter and energy at the smallest scales. For our book, *Volume II: The Codex*, the inclusion of the Schrödinger Equation is critical because it bridges the abstract mathematics of wave propagation, resonance, and harmonic structures with the architectural and cosmological frameworks we are exploring.

In our broader narrative, we have already introduced mathematical "structures", such as the Helmholtz and Poisson equations, which map out fields, waves, and potentials. The Schrödinger Equation extends these frameworks into the quantum domain, where the lattice of probability defines both limitation and possibility. It is in this interplay—between determinacy and indeterminacy—that our architecture of ascension takes root. Just as quantum superposition allows particles to exist in multiple states simultaneously, our terrestrial and cosmic architectures rely on overlapping grids of resonance and probability.

The Quantum Harmonic Oscillator: Quantized Resonance and Ascensional Lattices

The quantum harmonic oscillator (QHO) is one of the most fundamental and universal models in physics. It describes the behavior of a particle in a quadratic potential, revealing the quantized nature of resonance and vibration. Unlike the classical oscillator, whose motion is continuous, the QHO shows that oscillations occur in discrete energy levels with associated wavefunctions. This model is both mathematically elegant and conceptually profound, making it an essential foundation for Volume II of *The Codex*. It provides a rigorous physics-based counterpart to the symbolic themes of ascension, resonant lattices, and harmonic grids.

Mathematical Framework

The QHO potential is defined as:

$$V(x) = \tfrac{1}{2} m \omega^2 x^2$$

where m is the particle mass and ω the angular frequency. The Schrödinger equation:

$$(-\hbar^2/2m)\, d^2\psi/dx^2 + \tfrac{1}{2} m \omega^2 x^2 \psi = E \psi$$

yields solutions with quantized energies:

$$E_n = \hbar \omega (n + \tfrac{1}{2}), \quad n = 0,1,2,\ldots$$

and wavefunctions:

$$\psi_n(x) = N_n H_n(\xi)\, e^{(-\xi^2/2)}, \quad \xi = \sqrt{(m\omega/\hbar)}\, x$$

where H_n are Hermite polynomials and N_n a normalization factor. The result is a ladder of discrete vibrational states, each associated with a standing-wave probability distribution.

Resonance and Nodal Patterns

The wavefunctions $\psi_n(x)$ display nodal structures—points where the probability of finding the particle is zero. These nodal sets echo the resonance grids of the Helmholtz equation and the harmonic lines in architectural lattices. The higher the energy state, the greater the number of nodes, mirroring the way higher harmonic modes create increasingly intricate lattice structures.

Universality of the Harmonic Oscillator

The quantum harmonic oscillator is not an isolated case. Nearly every physical system near equilibrium behaves like a harmonic oscillator. This includes molecular vibrations, electromagnetic fields, solid-state lattices, and even cosmological perturbations. Thus, the QHO is the universal model of resonance across scales, from atomic to planetary to cosmic.

Symbolism and The Codex

For *Volume II: The Codex*, the QHO symbolizes the principle of ascension through discrete steps. Its energy spectrum is a ladder; each rung corresponds to a higher vibrational state. This reflects the Codex's narrative of ascensional architecture, in which one climbs through harmonic intervals rather than moving in a smooth continuum.

The nodal patterns of $\psi_n(x)$ embody the interplay of emptiness and fullness, silence and resonance, void and form. Just as sacred architecture uses modular intervals to encode spiritual ascent, the QHO provides the physics of quantized intervals as the foundation of vibration and structure.

Implications for Harmonic Architecture

By incorporating the QHO, *Volume II* situates harmonic grids within the deeper reality of quantization. Resonance is not only structural and planetary but also quantum-mechanical, with each step of ascension corresponding to a discrete leap in energy. This confirms that architecture, physics, and cosmology are linked through a universal ladder of resonance.

CHAPTER 10:

The Schrödinger Equation and Atomic Orbitals

The images presented are graphical solutions of the Schrödinger equation, one of the central equations of quantum mechanics. They represent the spatial forms of electron orbitals, the regions in space where an electron is most likely to be found in an atom. Each shape, color, and node arises directly from the mathematics of wavefunctions, making these images the most direct way to visualize the quantum structure of matter.

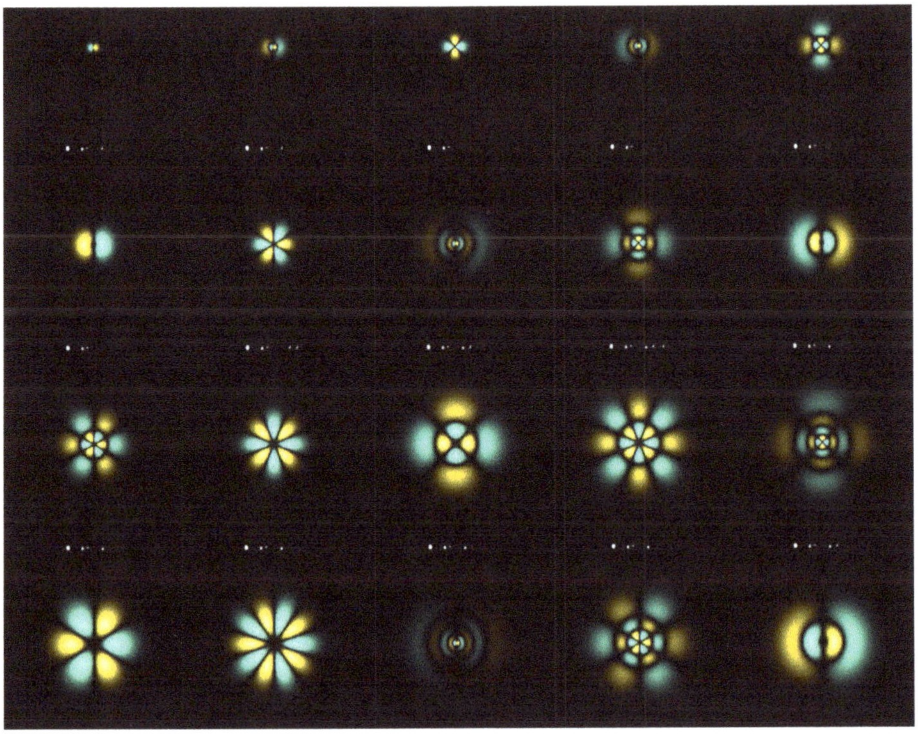

Figure 22 - Hydrogen Wave Function

Visualizations of electron orbitals derived from solutions to the time-independent Schrödinger equation. Each panel shows the spatial probability distribution (|ψ|2/\psi|^2|ψ|2) and phase (color-coded as positive/negative regions) of an orbital, defined by the quantum numbers nnn, lll, and mmm. The shapes illustrate how electron density organizes around the nucleus, with spherical s-orbitals, directional p-orbitals, and more complex d-orbitals. Nodes, where probability density is zero, appear as separations between colored regions, increasing in number with higher energy states.

Pauli Matrices within *The Codex* Framework Introduction: The Elemental Operators

Within *The Codex*, the universe is framed as a harmonic lattice, a tensorial grid through which resonance stabilizes both matter and cosmos,

Into this architecture we can insert the Pauli matrices, whose compact form embodies the irreducible symmetries of spin and the algebra of quantum states. As *The Codex* describes atomic void and planetary resonance as continuous aspects of one harmonic fabric, the Pauli matrices provide the algebraic scaffolding at the smallest scale, a minimal gateway into the architecture of fields

The Algebra of Symmetry

The Codex frequently invokes the continuity of lattices, fields, and resonances as stabilizers of both atoms and planetary structures

The Pauli matrices formalize this continuity at the microscopic level. Their commutation relations,

$$[\sigma x,\sigma y]=2i\sigma z,[\sigma y,\sigma z]=2i\sigma x,[\sigma z,\sigma x]=2i\sigma y.$$

mirror the tensorial dynamics of the resonance grid: rotations, couplings, and exchanges are not arbitrary but structured by algebraic invariants. Just as *The Codex* identifies spherical harmonics and contraction mappings as resonant forms on Earth's surface,

The Pauli matrices generate the rotational group SU(2), the universal symmetry of spinor fields.

Spin and the Bloch Sphere as Codex Microcosm

In *The Codex*, the Earth is portrayed as a planetary standing wave system, its sacred nodal points aligned with harmonic intersections.

The Bloch sphere, constructed from the Pauli matrices, is a microcosm of this planetary resonance. Each spinor state corresponds to a point on the sphere, every quantum superposition a wave node within a bounded geometry. The Bloch sphere becomes a miniature Codex lattice, an atomic Earth where states resonate across an algebraic grid.

Tensorial Extensions and Relativistic Fabric

The Codex emphasizes tensorial extensions, projecting local harmonics into global lattice structures.

Pauli matrices, embedded within the Dirac algebra, extend spin symmetry into relativistic space-time. This embedding mirrors the Codexian gesture of connecting atomic voids to planetary tensors: a single algebra, scaled outward, stabilizing multiple tiers of reality.

Philosophical Resonance

By placing the Pauli matrices inside *The Codex*, we see them not merely as mathematical operators but as elemental glyphs—archetypal forms encoding the same symmetry principles that structure lattices of sacred geography, resonant grids, and cosmogenic architectures. They are the algebraic seeds of the greater harmonic field. Where *The Codex* envisions Earth's tensor grid as a resonance of sacred nodes, the Pauli matrices provide the irreducible alphabet of symmetry, a universal script spoken at the smallest scales yet resonant through the largest.

Intuitive Geometric Significance of Pauli Matrices — A Designer's Summary

Based on Hongbing Zhang's paper, "Intuitive Geometric Significance of Pauli Matrices and Others in a Plane" (2017).

Overview

The paper reframes Pauli matrices (σx, σy, σz) and related operators as concrete geometric actions in a 2D real plane. By pairing simple matrix operations with familiar moves—mirroring across lines, 90-degree rotations, and axis projections—it makes the Lie algebra of SU(2) visually legible. The exposition then scales from 2D to special relativity via split-complex numbers, and finally to Dirac spinors and the gamma-matrix representation. The unifying theme is that algebraic structure can be represented as elementary geometric transformations.

Key Ideas and Takeaways

- Geometric actions in the plane: σx = mirror across y = x; $i\sigma y$ = 90-degree clockwise rotation; σz = mirror across x-axis.
- Pauli algebra by inspection: the commutators $[\sigma x, \sigma y] = 2i\sigma z$, $[\sigma y, \sigma z] = 2i\sigma x$, $[\sigma z, \sigma x] = 2i\sigma y$ emerge from composing those moves.
- Eigenvectors and eigenvalues: $\sigma z/2$ eigenvectors lie on axes; $\sigma x/2$ on the ±45-degree lines; $i\sigma y$ has no real-plane eigenvectors in this construction.
- Raising / lowering operators: $J+ = (\sigma x + i\sigma y)/2$ and $J- = (\sigma x - i\sigma y)/2$ act as exchange-projections between the axes, matching the textbook ladder behavior.
- Split-complex numbers and Lorentz transforms: writing a spacetime point as $S = t + x\sigma$ with $\sigma^2 = 1$ yields $e^{\sigma\theta} = \cosh\theta + \sigma\sinh\theta$, making boosts in hyperbolic rotations.
- From SU(2) to SO(3): exponentials $e^{i\sigma k\theta/2}$ act as plane rotations in "spinor space" and generate the same Lie algebra as real-space rotations Jk for SO(3).
- Spinors and Dirac equation: assembling (ψ, ϕ) with $E \to i\partial 0$ and $p \to -i\nabla$ leads to $(i\gamma^\mu\partial_\mu - m)\varphi = 0$ in the Weyl/Pauli block form.

Atomic Views: Macro–Micro Resonance

The following two diagrams are provided as companion visualizations. They echo the paper's central thesis by nesting rotational frames and orthographic symmetries. While Zhang's figures demonstrate the plane mechanics of Pauli actions, these spherical compositions show how stacked frames, axes, and great circles can scaffold multi-scale resonance, which is, useful when teaching the SU(2)↔SO(3) relation or spinor projections.

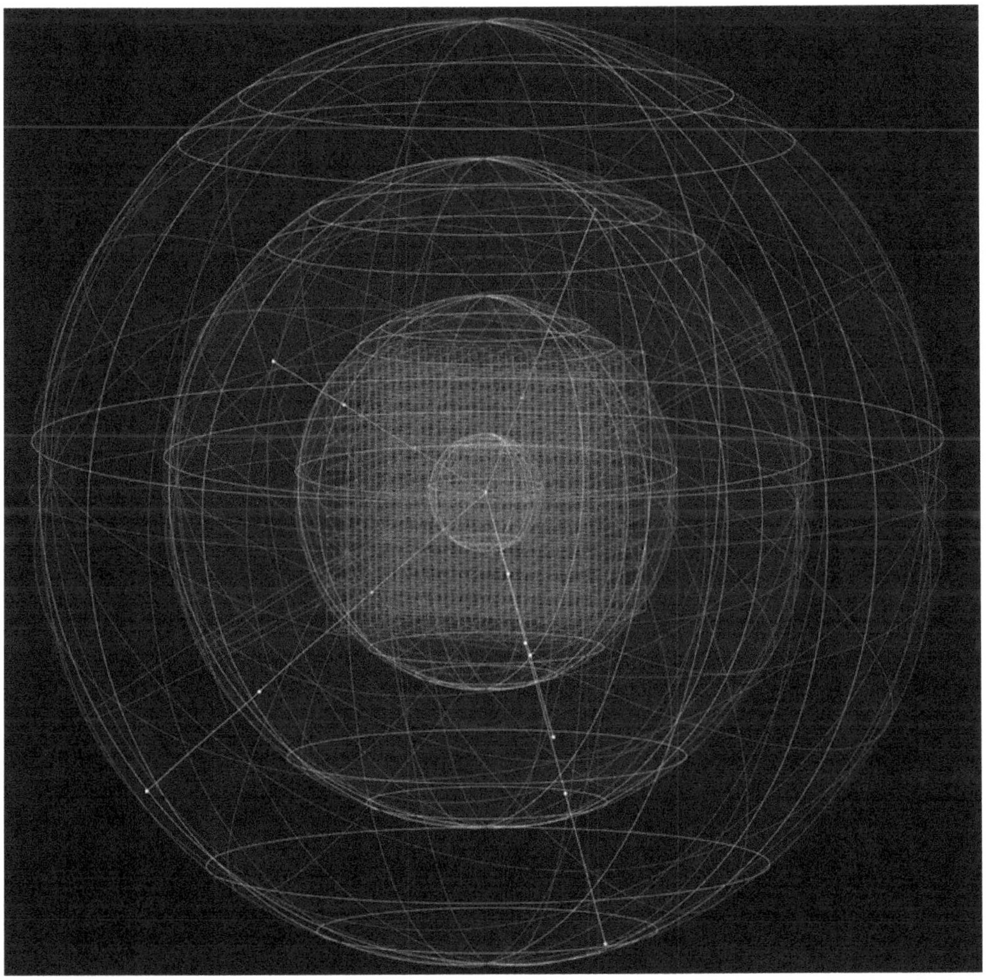

Figure 23 - Atomic View

Nested spheres and orthographic great circles; a spatial analog to commuting rotations and axis projections.

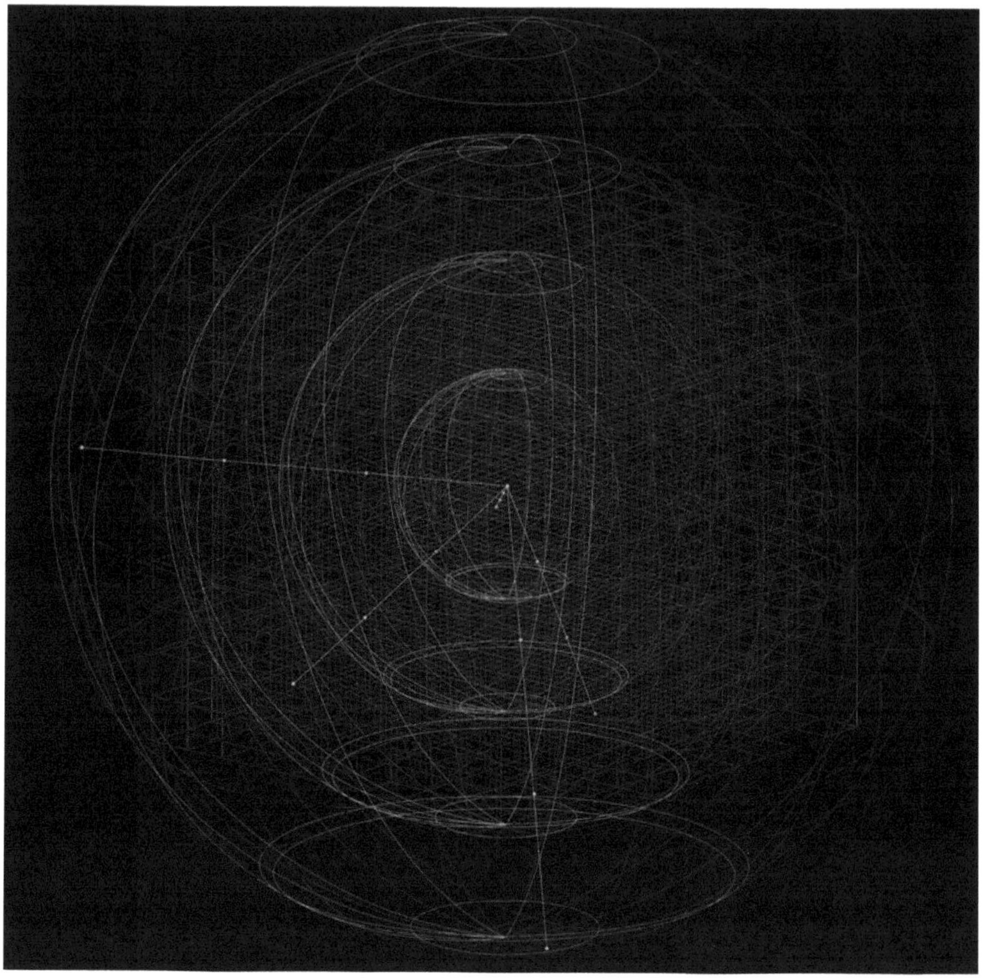

Figure 24 - Atomic View II

Layered spherical sections and axial stacks; useful for narrating ladder operations and projection sequences.

Why This Matters

For designers and educators, the paper is a blueprint for translating abstract algebra into first-principles geometry. It clarifies how a handful of 2×2 matrices encode mirrors, rotations, and projections—actions that scale from sketching in a plane, to composing boosts in spacetime, to building the Dirac operator. The approach is modular, visual, and immediately teachable.

Complex Cat States - A Theoretical Reading in Symbols

Twelve panels of interference patterns in phase space, read as a minimal mathematics of form.

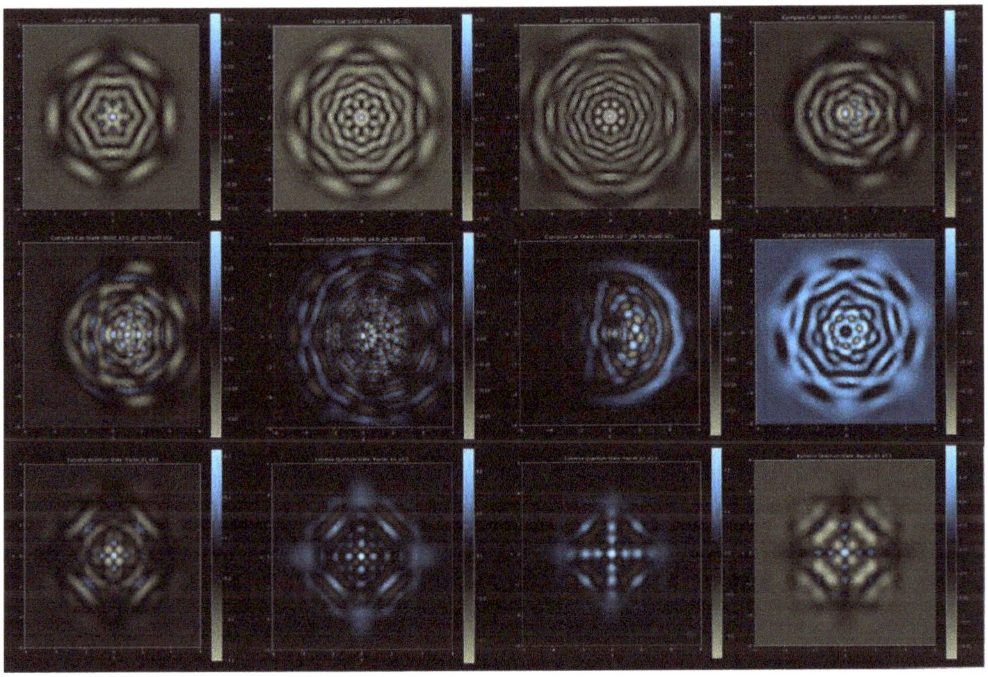

Figure 27 - Complex Cat States, Interference Atlas in Phase Space

Overview (what the grid shows)

A 3×4 matrix of square panels visualizes twelve Wigner functions W(q,p)W(q,p)W(q,p) on (q,p)∈[−5,5]2(q,p)\in[-5,5]^2(q,p)∈[−5,5]2 (axes marked in each panel). The diverging colormap encodes positive quasiprobability as warm beige/ivory and negative quasiprobability as cool navy/blue; near-zero values fade toward grey. Each title reads "Complex Cat State (…)" with parameters indicating rotational fold (symmetry order), displacement aaa (roughly the radius of the coherent-state constellation), and optional phase/rotation modifiers. A slim vertical color bar at the right of each panel reports the local range of WWW for that panel.

Each panel can be regarded as the visible trace of a hidden sentence written in superposition. Let the state be ψ, the plane be $\mathbb{R}^2$ with coordinates $x = (q, p)$, and the drawing be the quasiprobability $W\psi(x)$. What we see are level-sets of $W\psi$, where bright regions mark $W\psi(x) > 0$ and dark wells mark $W\psi(x) < 0$. The grammar is simple: symmetry + phase + distance → motif.

Minimal Model

Write the state as a finite superposition of coherent letters: $\psi = \Sigma_k a_k |\alpha_k\rangle$, with complex weights a_k and centers $\alpha_k \in \mathbb{C}$. What matters here is not the analytic form but three abstract controls:

- multiplicity $n = |\{k\}|$ (how many letters),
- arrangement α_k on a circle of radius r (constellation),
- phase profile $\varphi_k = \arg(a_k)$ (how the letters interfere).

Symmetry

If the constellation is cyclic, $\alpha_k = r \cdot e^{i(\theta_0 + 2\pi k/n)}$, then the underlying symmetry group is C_n. The pattern respects $W\psi(R_{2\pi/n}\, x) = W\psi(x)$ when the phases satisfy a compatible rhythm, e.g. $\varphi_{k+1} - \varphi_k = \text{const}$. Thus the rosettes are not ornaments but invariants of an action: rotations R_θ on the plane paired with a phase shift on the alphabet.

Interference Law

Superposition divides the page into three sets:

- Peaks $P = \{ x : W\psi(x) \gg 0 \}$,
- Wells $N = \{ x : W\psi(x) \ll 0 \}$,
- Nodal web $Z = \{ x : W\psi(x) = 0 \}$.

The eye follows Z as if it were a structural frame. Where phases reinforce ($\cos \Delta\varphi \approx 1$) we obtain rings and petals; where phases oppose ($\cos \Delta\varphi \approx -1$) the wells appear. The central checkerboard is the fixed point of all pairwise differences; it densifies as the constellation grows and the phases sharpen.

Parameters as Form-Makers

- Radius r controls radial spacing: larger r ⇒ more rings.
- Fold n controls azimuthal tessellation: n petals, n corridors of zeros.
- Global phase Φ rotates the interference lattice within the same symmetry class.
- Dephasing $0 \leq \lambda \leq 1$ contracts contrast by replacing $a\_k \mapsto (1-\lambda)a\_k$; visually, Z blurs and the wells lift toward 0.

Limits and Tendencies

Two asymptotic pictures organize the series:

(i) $r \to 0$: ψ behaves like a single letter; Wψ tends to a disk with soft halo; symmetry is latent.

(ii) $r \to \infty$ with n fixed: petals separate into a crown; the core becomes a tight checkerboard, an urban plan of interference.

Between these limits the atlas records phase-driven shears, ring splitting, and moiré corridors—each a readable response to the triplet (n, r, Φ).

Architectural Reading

Let Γ be the nodal framework Z. A designer may interpret Γ as a deployable grid: bays along rings, ribs along petals, programs in the wells (darks) and courtyards on peaks (lights). Material inversion—mapping sign(Wψ) to albedo or permeability—yields a literal nonclassical facade: the building is positive where interference is constructive and absorptive where it is destructive.

Closing Formula

The atlas says: form = orbit(ψ) under the joint action of rotations and phase shifts. Or, more explicitly,

$$W\psi = \mathfrak{W}(\Sigma_k a_k |\alpha_k\rangle) \Rightarrow \text{motif}(n, r, \Phi),$$

with $\mathfrak{W}$ the Wigner transform. The rest is editing: choose n for civic rhythm, r for depth, Φ for stance. In this sense the panels are studies of a single sentence rewritten in different meters.

Quantum States: an Atlas of Symmetry and Resonance

This plate reads like a field guide to standing-wave intelligence. Each tile stages a compact drama between constraint and possibility: a fixed boundary, a chosen fold-symmetry, and the free play of a complex amplitude settling into equilibrium. Petals open, rings pulse, stars rotate; nodes stitch the darkness; luminous corridors braid through a mesh of interference. What you see is a grammar of resonance—solutions to wave equations expressed as planforms—organized by n-fold rotational symmetry and by radial order. In the book's terms, this is a codex of modes: a discrete catalogue of how form precipitates from field.

Across the grid the same principle governs variety: symmetry selects spectrum. Choose a fold (3, 4, 5, …, 12), superpose equal-frequency plane waves at evenly spaced angles, and a characteristic rosette emerges. Increase radial order and concentric nodal rings appear; raise angular order and the rosette gains petals; combine them and moiré lattices bloom. Mathematically this lives in the solution families of the Helmholtz and Schrödinger equations on a disk: angular factors that behave like $e^{im\theta}$ (or their cosine–sine companions) and radial envelopes that echo Bessel behavior. Architecturally this becomes a rule: boundary + symmetry + frequency → tectonic motif.

This book has developed this rule as a through-line. Atomic orbitals, Chladni figures, harmonic oscillators, spherical harmonics, Green functions—each chapter turns the same key in a new lock and opens a chamber of patterns. The current image condenses that journey: every tile is a quantal room, a chamber of pressure and phase, whose walls are drawn by nodes. The bright filaments map ridges of probability (or intensity); the dark seams mark annulled presence where destructive interference seals the floor. In our design vocabulary, nodes are structural seams and circulation voids; antinodes are programmatic attractors and light wells; iso-contours become partitions, screens, trusses, and acoustic shells.

Read diagonally and a tectonic pedagogy unfolds. Along one diagonal, petal count climbs: an angular-momentum ladder—useful for rotating façades, radial framing, or turbine-like daylighting devices. Along another, radial nodes densify: a resonant compression—suggesting layered membranes, ring vaults, or ripple slabs that control acoustic or thermal flux. The tiles with crystalline starbursts point to quasi-periodic quiltings for ceilings and plazas; the foggier tiles, where rings smear into drift, suggest envelope modulations and graded porosities for passive ventilation and shading.

Scale the atlas up and the earlier chapters on the planetary tensor grid reappear. Replace the disk by the sphere, lift these planar rosettes into spherical harmonics, and the same logic organizes the Earth's standing-wave shells—our language for ley-line lattices, nodal knots, and resonant basins. Sacred geographies, civic networks, and ecological corridors become global eigenmodes: intersections of dominant harmonics where temples, markets, and habitats anchor. Scale the atlas down and the quantum essays return: localized wells, tunneling channels, and band structures, each a micro-tectonic version of the same symphonic law.

- Matrix framework: treat each tile as a parameter preset (n, m, k). Vary symmetry n, angular order m, and radial index; map iso-contours to walls; map gradient flow to circulation; map nodal rings to courtyards and oculi.
- Material translation: print antinodal ridges as denser ribs; assign high-node zones to apertures; use phase to alternate material (glass ↔ lattice), enabling daylight caustics analogous to the image's bright filaments.
- Environmental coupling: tune the pattern's frequency to acoustic targets (reverberation time), thermal targets (shading beats), or structural eigenfrequencies (vibration control). A plan ceases to be a diagram and becomes a harmonic device.

Philosophically, the plate advances this book's thesis: form is organized energy. Modes are the unit steps of that organization, and symmetry is the contract that lets matter sing. From atom to temple to terrain, the same spectral alphabet writes space. The atlas of tiles is therefore both ornament and oracle: a catalogue of beautiful answers the field gives when asked precise questions.

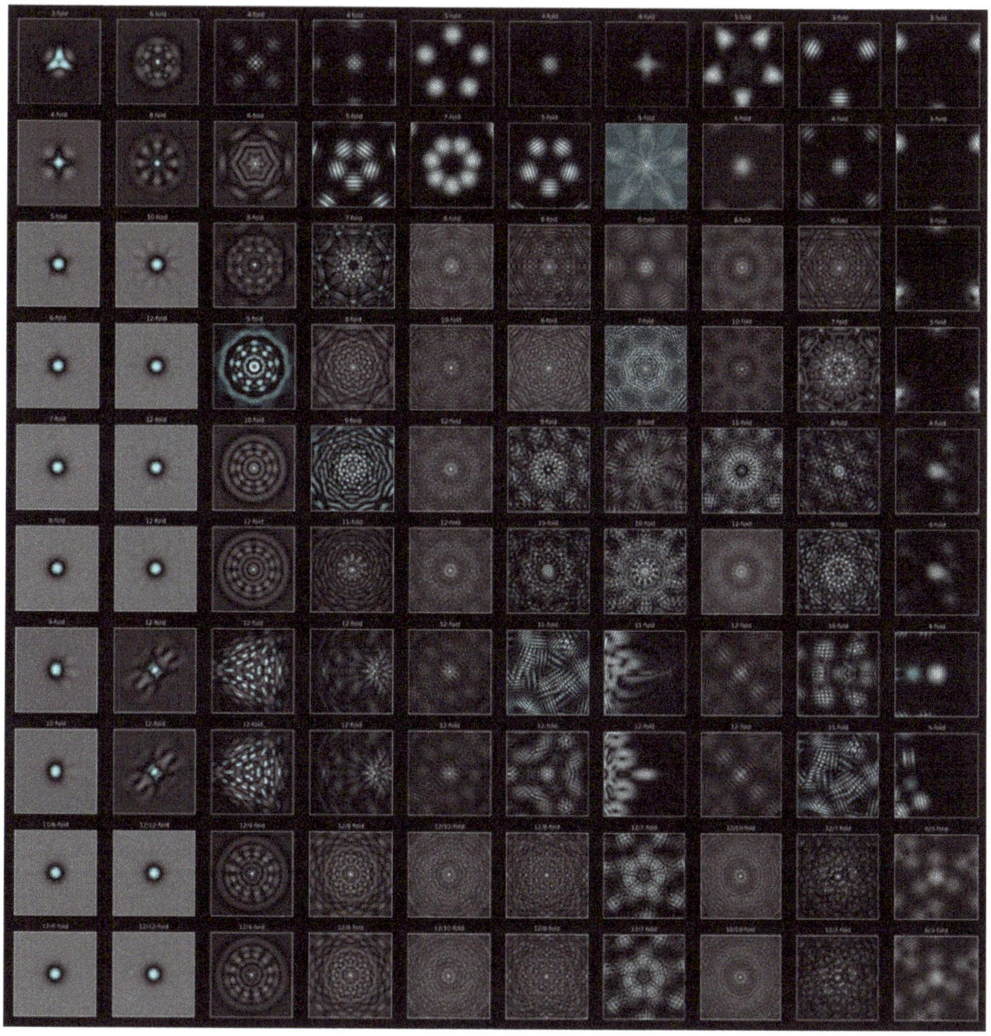

Figure 28 - Quantum States, n-Fold Atlas

A matrix of standing-wave planforms generated by equal-frequency superpositions with n-fold rotational symmetry. Columns enumerate angular orders and radial nodes; bright filaments trace antinodes (programmatic attractors), dark seams trace nodes (structural seams and voids). The plate serves as a parametric library for translating resonance into architectural tectonics—scalable from quantum chambers to planetary lattices.

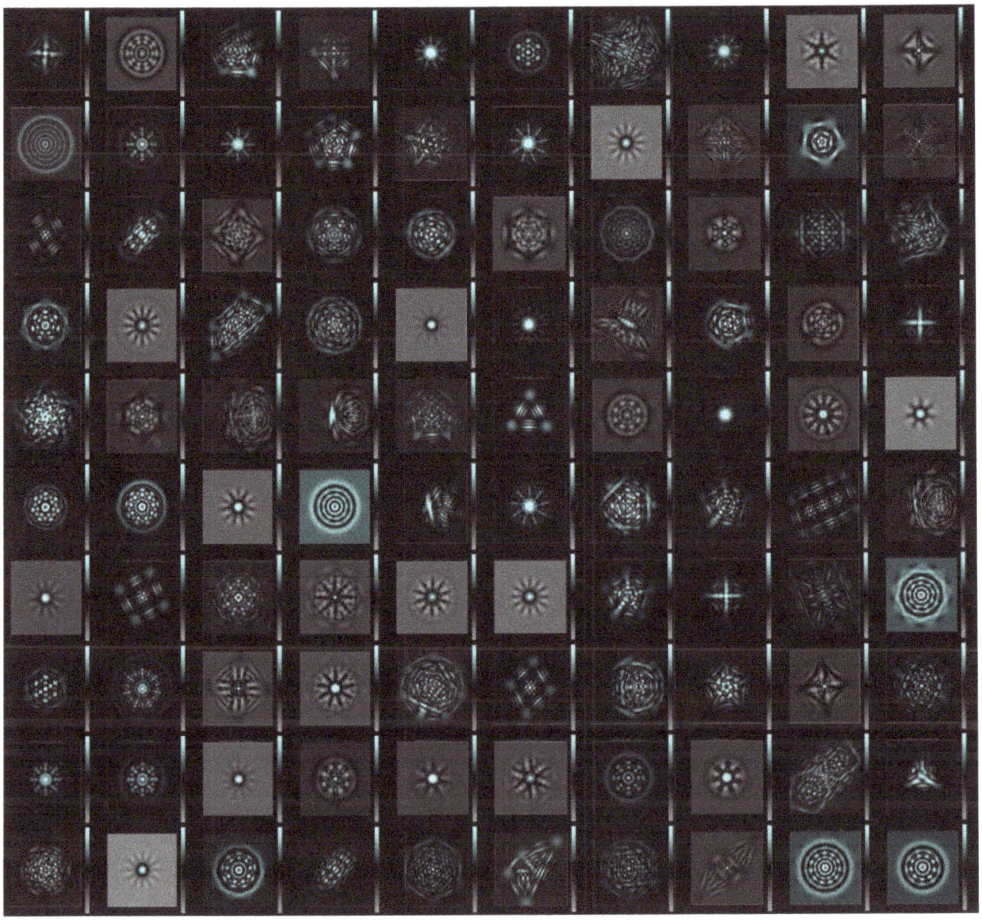

Figure 29 - The Quantum Atlas (V2)

Matrix of standing-wave planforms arranged by symmetry class and radial order. Bright filaments trace antinodes; dark seams trace nodal corridors. The plate functions as a parametric library for translating resonance into tectonics.

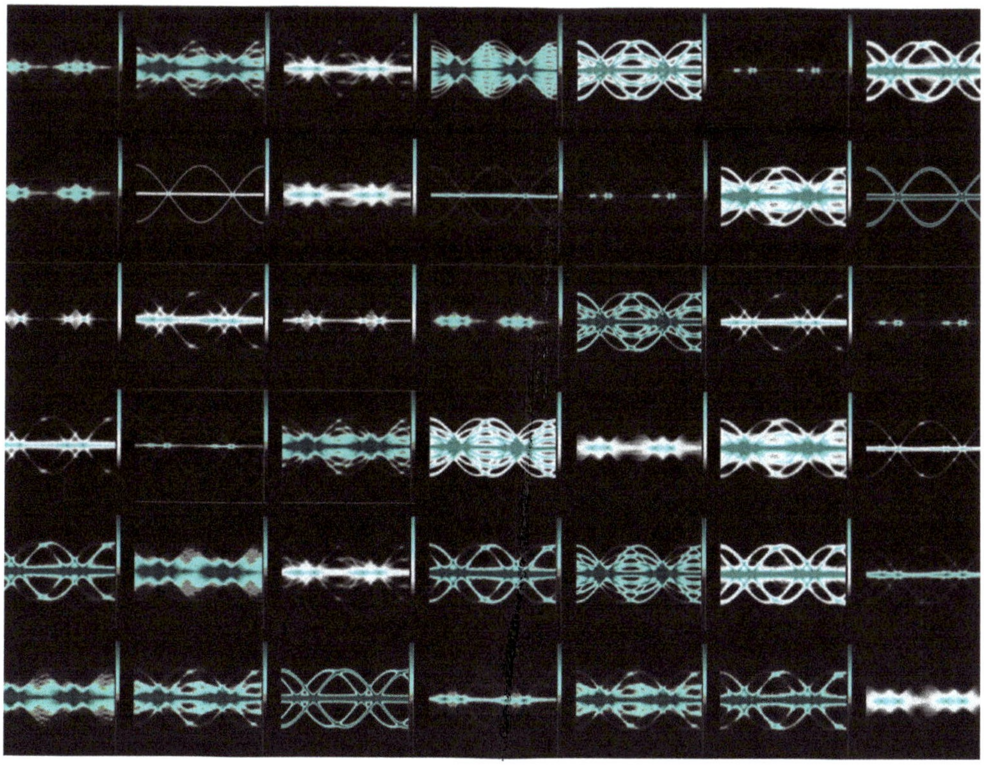

Figure 30 - Matrix Quantum Wave Band Plate

Each panel shows a bounded 1D superposition rendered as luminous antinodal ridges and nodal lanes. Coupling, beating, and damping generate the braid-like variations across the grid; these map directly to ribs, joints, and envelope modulations in linear architecture.

Orientation

This plate assembles a family of one-dimensional standing-wave fields into a matrix of narrow horizontal chambers. Each chamber records a superposition of a few harmonics in a bounded guide, rendered as luminous turquoise ridges on a black background. Across the matrix the ridges couple, separate, braid, and flatten: a didactic filmstrip of dispersion and interference. What we read here is a band language — the longitudinal counterpart to the Quantum Atlas — a spectral storyboard for corridors, bridges, galleries, and pipes.

How to Read the Bands

Bright ridges = antinodes (energy or flux concentration); dark lanes = nodes (null flux).

Thin, sharp ridges indicate coherent superpositions; thicker ribbons indicate dephasing or envelope modulation.

Pairs of ridges that approach and repel narrate avoided crossings — the architectural analogue of coupled modes in a guide.

Period doubling along the corridor marks a higher harmonic index; envelope beats mark interference between close frequencies.

From Bands to Tectonics

Translate bands directly into section-elevations of long spaces:

- Antinodal ridges → continuous ribs, cable runs, or light shelves that carry force, air, or photons.
- Nodal lanes → relief joints, service chases, circulation clearances, or acoustic baffles.
- Beats/envelopes → programmatic cadence: galleries and alcoves repeating with controlled variation.
- Avoided crossings → branch–merge moments in bridges and corridors, stabilized by coupling rings.

Because bands are one–dimensional spectra, they excel at scripting procession. A visitor moves with the phase; the program sits at antinodes; services align to nodes.

Environmental Coupling

Acoustics: choose band spacing to keep speech in antinodal clarity while steering echoes into nodal absorbers.

Light: use ridge-aligned louvers and micro-mirrors to cast linear caustics; modulate period for circadian pacing.

Thermal/air: treat antinodes as laminar corridors; use nodal zones as low-pressure returns; tune envelope beats to seasonal regimes.

Structure: align continuous ribs with maximal curvature of the band field to minimize material for a given stiffness.

Protocols

Protocol A — Corridor Synthesis: select L and three to five target frequencies; extract iso-lines for ribs (antinodes) and joints (nodes); introduce a weak coupling region to create an avoided-crossing zone; add damping to thicken texture in service bays.

Protocol B — Bridge/Beam Morph: map band peaks to variable-depth web stiffeners; map nodes to perforations; phase-advance the pattern along the span to script experience and distribute mass.

Relation to the Quantum Atlas

Planar rosettes are 2D eigenpictures; band plates are 1D eigenfilms. Together they complete a kit: rooms from rosettes, routes from bands. Lift the bands to the sphere and they become great-circle waveguides — meridian lattices in the planetary tensor grid. Drop them to photonic scale and they are waveguides and gratings. Scale is merely a tuning knob; the language stays spectral.

Symbols

$$(\nabla^2 + k^2)\,\psi = 0 \quad \text{— Helmholtz standing-wave condition.}$$

$$\psi(r,\theta) = R_m(r)\, e^{i m \theta}, \quad R_m \sim J_m(k r).$$

$$[A,B] = AB - BA; \quad \{A,B\} = AB + BA.$$

$$d\rho/dt = -i[H,\rho] + \Sigma_k \left(L_k \rho L_k^\dagger - \tfrac{1}{2}\{L_k^\dagger L_k, \rho\} \right).$$

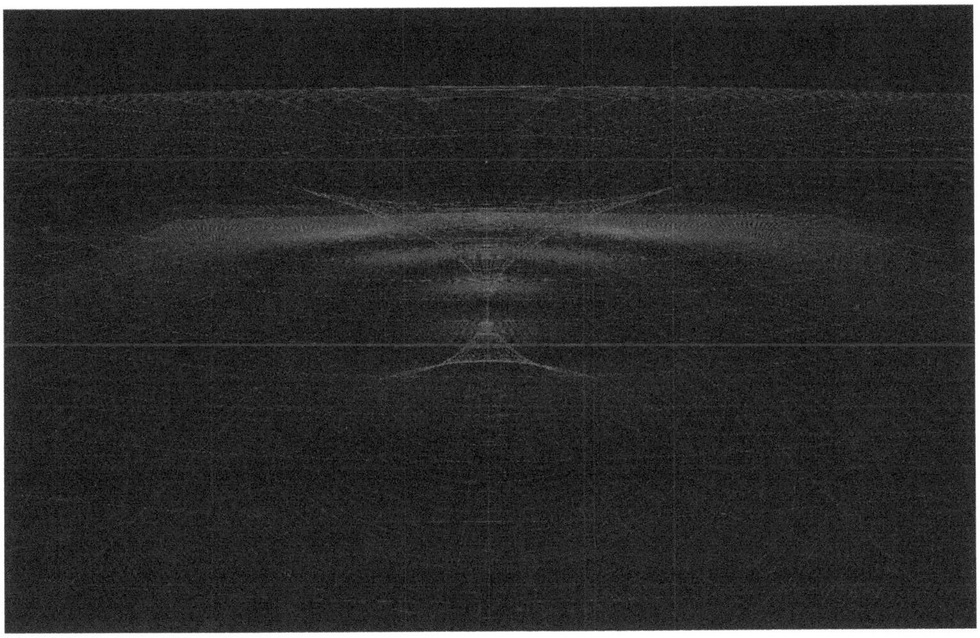

Figure 31 - Minkowski light-cone scaffold over scalar-wave interference
a conceptual 'gravitational matrix.'

CHAPTER 11:

A Quantum–Relativistic Atlas - Synthesizing Einstein's 1905 Quartet with Schrödinger's 'My View of the World'

Introduction

This chapter offers a single navigational chart—an atlas—where two revolutions meet. Einstein's 1905 papers established the quantum of light, the statistical reality of atoms via Brownian motion, and the kinematics and energetics of spacetime; Schrödinger, reflecting in *My View of the World,* connected the formalism of wave mechanics to a philosophical stance about unity of mind and world. The figure that ends the previous chapter overlays a light-cone lattice on a rippling interference field, hinting at how causal structure (relativity) constrains and shapes lawful patterns of appearance (quantum waves). What follows is a synthesis: concise statements of core results and equations, and an interpretive bridge between causal cones and wave amplitudes.

Light Quanta and the Photoelectric Effect

Einstein posited that radiation of frequency ν is quantized in energy packets $E = h\nu$. For electrons ejected from a metal with work function φ, the maximum kinetic energy is $K\_max = h\nu - \varphi$. This linear relation isolates Planck's constant in experiment and discloses particulate aspects of light.

$$E = h\,\nu$$

$$K\_max = h\,\nu - \varphi$$

Brownian Motion and the Molecular Reality

By modeling suspended particles buffeted by molecular collisions, Einstein derived a diffusion law that permits determination of Avogadro's number. The Stokes–Einstein relation connects the diffusion coefficient

D to temperature, viscosity, and particle radius r. Mean-squared displacement grows linearly with time.

$$D = k\_B T / (6 \pi \eta r)$$

$$\langle x^2 \rangle = 2 D t \quad (1D)$$

$$\langle r^2 \rangle = 6 D t \quad (3D)$$

On the Electrodynamics of Moving Bodies (Special Relativity)

Assuming (i) the laws of physics are invariant in all inertial frames and (ii) the speed of light is constant, Einstein obtained the Lorentz transformations, the relativity of simultaneity, time dilation, and length contraction. The invariant interval and proper time bind kinematics to causality and define the light-cone structure.

$$\gamma = 1 / \mathrm{sqrt}(1 - v^2/c^2)$$

$$x' = \gamma (x - v t), \; y' = y, \; z' = z$$

$$t' = \gamma (t - v x / c^2)$$

$$\Delta t' = \gamma \Delta t \quad \text{(for events at same place in S)}$$

$$L = L_0 / \gamma \quad \text{(length contraction)}$$

$$s^2 = -c^2 t^2 + x^2 + y^2 + z^2 \quad \text{(Minkowski interval)}$$

$$d\tau = dt \sqrt{(1 - v^2/c^2)} \quad \text{(proper time)}$$

$$u' = (u - v) / (1 - u v / c^2) \quad \text{(1D velocity addition)}$$

Does the Inertia of a Body Depend Upon Its Energy Content?

A change in a body's energy ΔE entails a change in inertial mass $\Delta m = \Delta E/c^2$. The general relation between energy, momentum, and rest mass consolidates dynamics across frames.

$$E = m c^2$$

$$E^2 = (p c)^2 + (m c^2)^2$$

Schrödinger's *My View of the World* and the Wave Picture

Schrödinger's philosophical essays stress a unitary view: the subject-object split is provisional; the world is one. His physics, wave mechanics, encodes phenomena in a complex amplitude ψ whose squared magnitude gives probabilities. The evolution of ψ is governed by the Schrödinger equation—the generator of lawful becoming in the quantum atlas.

Core Equations of Wave Mechanics

$$i \hbar \, \partial\psi/\partial t = \hat{H} \psi \quad \text{(time-dependent Schrödinger equation)}$$

$$-(\hbar^2 / 2m) \nabla^2 \psi + V \psi = E \psi \quad \text{(time-independent Schrödinger equation)}$$

$$\rho = |\psi|^2, \ \partial\rho/\partial t + \nabla\cdot j = 0, \ j = (\hbar / 2mi) (\psi^* \nabla\psi - \psi \nabla\psi^*)$$

$$\hat{p} = -i \hbar \nabla, \ \hat{H} = \hat{p}^2/(2m) + V(x)$$

$$[\hat{x}\_i, \hat{p}\_j] = i \hbar \delta\_\{ij\}, \ \Delta x \, \Delta p \geq \hbar / 2$$

$$\langle A \rangle = (i/\hbar) \langle [\hat{H}, \hat{A}] \rangle + \langle \partial\hat{A}/\partial t \rangle \quad \text{(Ehrenfest theorem)}$$

$$\lambda = h / p \quad \text{(de Broglie relation)}$$

Bridge: Cones, Waves, and the Metric of Appearance

The light cone carves spacetime into three causal classes: timelike, spacelike, and null. Quantum amplitudes must propagate within these causal constraints. In relativistic wave equations, the d'Alembert operator $\Box$ encodes Minkowski geometry; null characteristics of $\Box$ generate the cone evident in the figure. A scalar quantum with mass m obeys the Klein–Gordon equation; massless quanta satisfy the wave equation $\Box\psi = 0$. These equations reconcile phase-coherent spread with causal reach.

$$\Box \equiv -(1/c^2)\, \partial^2/\partial t^2 + \nabla^2$$

$$(\Box + (m\, c\, /\, \hbar)^2)\, \psi = 0 \quad \text{(Klein–Gordon)}$$

$$\Box\, \psi = 0 \quad \text{(massless wave/field)}$$

$$F_{\mu\nu} = \partial_\mu A_\nu - \partial_\nu A_\mu,\ \ \partial_\mu F^{\mu\nu} = \mu_0 J^{\nu} \quad \text{(Maxwell in covariant form)}$$

$$ds^2 = g_{\mu\nu}\, dx^{\mu}\, dx^{\nu} \quad \text{(metric form; here } g_{\mu\nu} = \eta_{\mu\nu} \text{ for flat spacetime)}$$

Interpretation and Architectural Analogy

The atlas metaphor emphasizes coordination. Einstein contributes the frame geometry and energetic budget; Schrödinger supplies the wave-amplitude grammar that populates those frames. Causal cones delimit admissible influence; quantum phases imprint interference textures within that delimitation. In design terms, the cone lattice is the structural grid; the wavefield is the occupancy pattern. Unity—central to Schrödinger's philosophy—emerges when we read both layers together: probability as a pattern of appearance within a metric that governs what may affect what and when.

Equation Compendium (for quick reference)

- Planck–Einstein: $E = h\nu$; Photoelectric: $K\_max = h\nu - \varphi$
- Brownian motion: $D = k\_B T / (6 \pi \eta r)$; $\langle x^2 \rangle = 2 D t$; $\langle r^2 \rangle = 6 D t$
- Lorentz factor: $\gamma = 1/\sqrt{(1 - v^2/c^2)}$; Lorentz transforms: $x' = \gamma(x - vt)$, $t' = \gamma(t - v x / c^2)$
- Interval & proper time: $s^2 = -c^2 t^2 + x^2 + y^2 + z^2$; $d\tau = dt \sqrt{(1 - v^2/c^2)}$
- Velocity addition (1D): $u' = (u - v)/(1 - u v / c^2)$
- Mass–energy: $E=mc^2$; Energy–momentum: $E^2 = (p c)^2 + (m c^2)^2$
- Schrödinger (TD/indep.): $i\hbar\, \partial\psi/\partial t = \hat{H}\psi$; $-(\hbar^2/2m)\nabla^2\psi + V\psi = E\psi$
- Probability current: $\partial\rho/\partial t + \nabla\cdot j = 0$, $j = (\hbar/2mi)(\psi^*\nabla\psi - \psi\nabla\psi^*)$
- Operators/commutator: $\hat{p} = -i\hbar\nabla$; $[\hat{x}\_i, \hat{p}\_j] = i\hbar\, \delta\_\{ij\}$; $\Delta x\, \Delta p \geq \hbar/2$
- Relativistic waves: $\Box \equiv -(1/c^2)\partial^2/\partial t^2 + \nabla^2$; $(\Box + (m c / \hbar)^2)\, \psi = 0$; $\Box\psi = 0$
- Maxwell (covariant): $F\_\{\mu\nu\} = \partial\_\mu A\_\nu - \partial\_\nu A\_\mu$, $\partial\_\mu F\^\{\mu\nu\} = \mu_0 J\^\{\nu\}$

CHAPTER 12:

Coupled Dynamical Systems on a Constrained Manifold: A Swarm-Wave-Projection Framework

The simulation implemented in this code is not merely a numerical experiment; it is a geometric field theory discretized into agents. Its governing logic may be read as a coupled dynamical system evolving on a constrained manifold, rendered through a conformal projection that transforms motion into ornament. The code formalizes a universe where swarm intelligence, quantum probability, and projection geometry are not metaphors but operators.

State Space and Manifold Constraint

Each particle is represented by a state vector

xi(*t*)∈*R*3,*vi*(*t*)∈*R*3xi(t)∈R3,vi(t)∈R3

with a hard constraint enforcing motion on the 2-sphere:

∥*xi*∥=*R*∥xi∥=R

This is achieved numerically by projection after each integration step, ensuring that all dynamics remain tangential. The sphere functions as a compact phase domain—analogous to a closed universe—while allowing local complexity without boundary artifacts.

Swarm Dynamics as Local Mean-Field Approximation

The classical boids model is reformulated in a vectorized, grid-aggregated mean-field form. For each particle *i*i, acceleration derives from three neighborhood operators:

dxidt=*vi*dtdxi=vi*dvidt*=*αAi*+*βCi*−*γSi*dtdvi=αAi+βCi−γSi

where the terms are defined as:

Ai=⟨*v*⟩*Ni*−*vi*Ai=⟨v⟩Ni−vi*Ci*=⟨*x*⟩*Ni*−*xi*Ci=⟨x⟩Ni−xi*Si*=∑*j*∈*Nixi*−*xj*∥*xi*−*xj*∥2Si=j∈Ni∑∥xi−xj∥2xi−xj

The crucial mathematical shift is that neighborhoods are not enumerated pairwise; instead, particles sample cell-wise averages computed in $O(N)\mathrm{O(N)}$ time. This converts a classical $N2\mathrm{N2}$ interaction into a coarse-grained statistical field, aligning the swarm with continuum mechanics rather than discrete mechanics.

Quantum-Inspired Wavefunction Field

Embedded within the swarm is a time-evolving complex scalar field:

$$\psi(x,y,t)=\sum k=1KAk\exp(i(kk\cdot r-\omega kt+\phi k))\exp(-\|r-rk\|22\sigma k2)$$
$$\mathrm{\psi(x,y,t)=k=1\sum KAkexp(i(kk\cdot r-\omega kt+\phi k))exp(-2\sigma k2\|r-rk\|2)}$$

This field is not symbolic—it is explicitly computed on a lattice and sampled by particles. Two derived quantities drive motion:

Probability density

$$\rho(x,y,t)=|\psi(x,y,t)|2\mathrm{\rho(x,y,t)=|\psi(x,y,t)|2}$$

Phase

$$\theta(x,y,t)=\arg(\psi)\mathrm{\theta(x,y,t)=arg(\psi)}$$

Particles experience forces proportional to the gradients of these quantities:

$$F\rho=\nabla\rho F\theta=\nabla\theta\mathrm{F\rho=\nabla\rho F\theta=\nabla\theta}$$

Crucially, these gradients are **lifted into the tangent space of the sphere**, yielding:

$$F\mathrm{wave}=\eta 1\,PT(\nabla\rho)+\eta 2\,PT(\nabla\theta)\mathrm{Fwave=\eta 1PT(\nabla\rho)+\eta 2PT(\nabla\theta)}$$

where $PT\mathrm{PT}$ is the projection onto the local tangent plane. This aligns the simulation with Madelung hydrodynamics: particles follow probability flow rather than forces from other particles.

Stereographic Projection as Perceptual Operator

Rendering applies a conformal mapping from the sphere $S2S2$ to the plane:

$$\pi{:}S2\backslash\{N\}\to R2\pi{:}\mathrm{S2}\backslash\{\mathrm{N}\}\to\mathrm{R2}(x,y,z)\mapsto(xR-z,yR-z)(\mathrm{x,y,z})\mapsto(\mathrm{R-zx,R-zy})$$

This projection preserves angles but distorts area:

$$\det(D\pi)\propto 1(R-z)2\det(\mathrm{D}\pi)\propto(\mathrm{R-z})21$$

Thus, density increases dramatically near the projection pole, producing luminous cores and radial filaments. Importantly, nothing special occurs dynamically at these points—the singularity is epistemic, not physical. The image reveals how geometry itself manufactures apparent structure.

Criticality Parameter and Phase Transition

A scalar control parameter $\lambda\in[0,1]\lambda\in[0,1]$ modulates the entire system:

$$\alpha(\lambda),\beta(\lambda),\gamma(\lambda),\eta1(\lambda),\eta2(\lambda)\alpha(\lambda),\beta(\lambda),\gamma(\lambda),\eta1(\lambda),\eta2(\lambda)$$

As $\lambda\to1\lambda\to1$, the system undergoes a qualitative transition:

smooth flow ⟶ filamentation ⟶ braided critical structuressmooth flow ⟶ filamentation ⟶ braided critical structures

This is mathematically analogous to near-critical turbulence: alignment increases, damping decreases, and higher-frequency wave modes activate. The simulation therefore lives perpetually near a self-organized critical state.

Integration Scheme and Stability

Time evolution follows a semi-implicit Euler method:

$$vt+\Delta t=\mu vt+\Delta t\,F\mathrm{vt}+\Delta\mathrm{t}=\mu\mathrm{vt}+\Delta\mathrm{tF}xt+\Delta t=xt+\Delta t\,vt+\Delta t\mathrm{xt}+\Delta\mathrm{t}=\mathrm{xt}+\Delta\mathrm{tvt}+\Delta\mathrm{t}$$

with immediate projection enforcing:

$$xt+\Delta t \leftarrow Rxt+\Delta t \| xt+\Delta t \| \mathrm{xt+\Delta t \leftarrow R \| xt+\Delta t \| xt+\Delta t}$$

This produces a numerically stable flow on a curved manifold without requiring geodesic integration.

Interpretation

Mathematically this code defines:

- A measure-preserving flow on $S2$S2
- Driven by a complex-valued scalar field
- Observed through a conformal projection
- Stabilized near criticality

What emerges visually—mandalas, armadas, murmuration lattices—is not encoded explicitly. It is the fixed point of interacting operators.

Structure=limit→∞(Swarm∘Wave∘Projection)Structure=t→∞lim (Swarm∘Wave∘Projection)

In this sense, the code is not simulating objects.

It is simulating relations, and the image is the trace those relations leave on perception.

Figure 32 - Stereographic Murmurations

A sequence of emergent symmetry states arising from the coupled swarm–wave dynamics defined in Universe.py. Each panel represents a momentary phase portrait of a measure-preserving flow on the sphere S2S^2S2, rendered via stereographic projection. Luminous nodal cores and radiating filaments correspond not to material concentrations but to regions of amplified probability density and phase alignment, intensified by the conformal distortion of the projection near the pole. The repeated dihedral and radial symmetries are the visible trace of interacting operators, local mean-field swarm alignment, quantum-inspired probability flow, and spherical constraint, driven near criticality. The image should be read as ornament generated by motion: a cartography of relations rather than objects.

CHAPTER 13:

Crystalline Megastructures and Architectural Ascension

This chapter extends the codex into the architectural domain, grounding it in rigorous mathematics, tensorial analysis, and historical examples. Having traced the continuity from atomic void through the planetary tensor grid, we now demonstrate how human architecture materializes these structures into crystalline megastructures. Architecture is not merely functional; it is the resonant crystallization of planetary harmonics, aligned with nodal points of the Earth's morphogenetic grid. We establish theorems that prove architecture as a contraction operator, provide centered formal proofs, and show how sacred geometry encodes tensorial invariants. The chapter concludes by situating megastructures as ascension channels: crystalline operators bridging void, planet, and consciousness.

Invocation & Continuity

The atom revealed emptiness structured by probability. The planet revealed resonance structured by harmonic grids. Architecture now reveals form structured by ascension. The lineage is:

$$\text{Void} \rightarrow \text{Relation} \rightarrow \text{Resonance} \rightarrow \text{Form}$$

Just as geodesics concentrate energy at nodal points, so too civilizations concentrated stone, light, and geometry into crystalline order. These megastructures did not merely decorate the landscape—they anchored resonance into permanence. Architecture is the continuation of cosmic law into cultural form.

Tensorial Grammar of Architecture

Architecture can be described as a tensorial operator on the planetary grid. Let Ψ^* be the planetary equilibrium state. Then architecture acts as a stabilizing contraction mapping the harmonic field into crystalline form:

$$\mathcal{A} : \Psi\text{^*} \rightarrow \Psi\text{^*} \otimes \mathbb{C}\text{^n}$$

where n encodes crystalline symmetries (e.g., tetrahedral, octahedral, dodecahedral). The architectural tensor preserves invariants:

$$d\_X(\mathcal{A}u, \mathcal{A}v) \leq \Lambda\, d\_X(u,v), \quad \Lambda < 1$$

Thus, architecture contracts resonance into matter, preventing dispersion, ensuring continuity, and stabilizing planetary energy.

Sacred Geometry as Projection of Harmonics

Spherical harmonics project into planar tilings and polyhedral lattices. The Pyramids of Giza align with $l = 2$ quadrupole harmonics; Stonehenge aligns to solar azimuth harmonics; Mount Kailash resonates with $l = 3$ octupole forms. We define the geometric projection operator:

$$\Pi\_\{geo\} : Y\_\ell\text{^}m \rightarrow P\_\ell\text{^}m \otimes G\_n$$

where G_n are Platonic or Archimedean solids. Sacred architecture thus represents a geometric contraction of spherical harmonics into crystalline volumes. Geometry is resonance rendered tangible.

Crystalline Lattices as Ascension Operators

Every crystalline lattice has a point group symmetry $G \subset SO(3)$. The architectural ascension operator C is defined as:

$$\mathfrak{C} : (\Psi\text{^*}, G) \rightarrow \Omega\_\{asc\}$$

where Ω_{asc} is the space of ascended configurations: pyramids, mandalas, ziggurats, gothic cathedrals.

These structures perform three roles:

1. Anchor — stabilize planetary resonance at nodal sites.

2. Amplify — increase coherence of morphogenetic fields.

3. Ascend — serve as conduits between planetary tensor grids and cosmic resonance.

Proof of Architectural Contraction

We now prove that architecture functions as a contraction operator.

Lemma 1 (Symmetry Fixation): Crystalline point groups act as attractors on local fields:

$$\|F_{n+1} - F_n\| \leq (1-\kappa_G) \|F_n - F_{n-1}\|, \quad 0<\kappa_G<1$$

Lemma 2 (Resonant Amplification): Polyhedral volumes concentrate harmonic energy:

$$\int_{V_n} |\psi|^2 \, dV \geq C \int_{\partial V_n} |\psi|^2 \, dA$$

Proposition: The architectural operator is contractive:

$$\mathcal{A} = \mathfrak{C} \circ \Pi_{geo} \circ \Psi^*$$

Theorem (Architectural Fixed Point): There exists a unique equilibrium Φ^* such that:

$$\mathcal{A}(\Phi^*) = \Phi^*$$

Interpretation: Architectural ascension is mathematically inevitable where resonance and human intention intersect.

Crystalline Megastructures Across Cultures

- Giza Pyramids (Egypt): Quadrupole harmonic contraction into tetrahedral form.
- Stonehenge (Britain): Circular projection of solar-lunar harmonics into lithic resonance.
- Borobudur (Indonesia): Fractal stupa array encoding ascending harmonics.
- Chartres Cathedral (France): Gothic crystalline light-geometry embedding rose window harmonics.
- Teotihuacan (Mexico): Planetary alignment of pyramids reflecting Venus resonance cycles.

Each exemplifies the principle: crystalline megastructure =

contraction of planetary harmonic fabric into human form.

Corollary: Architecture as Ascension Channel

Architecture is not passive. It transduces planetary energy into consciousness. Standing within these megastructures alters the observer's state, aligning neural resonance with planetary grids. Formally, consciousness operator C (Ch. 8) is modified by architectural mediation:

$$C' = \mathcal{A} \circ C$$

Thus, architectural space is itself an operator of ascension.

CHAPTER 14:

The Architecture of Ascension

Introduction: Architecture Beyond the Physical

The Architecture of Ascension transcends material construction, positioning architecture as a cosmological operator. These forms are not inert artifacts, but resonant instruments that integrate the vibrational strata of the cosmos, Earth, and human consciousness. At its foundation lies the fractal megastructure—an architectural archetype that expands and contracts across scales, echoing the recursive self-similarity found in nature and encoded in cosmic law.

The ascension paradigm situates architecture as a fixed-point attractor in the planetary resonance field, stabilizing the convergence of higher-dimensional energies into habitable form. In this view, buildings, cities, and landscapes function not merely as spaces of shelter, but as coherence operators—local condensations of universal harmonic order.

Fractals as Universal Blueprint

Fractals emerge wherever self-similarity and recursive generation converge. They form the grammar of the cosmos present in river deltas, vascular networks, lightning branches, and galactic arms. In ascension architecture, fractals embody the law of infinite connectivity between matter, energy, and consciousness.

Mathematically, the fractal dimension is given by:

$$D = \log(N) / \log(1/r)$$

Algorithmic Design as Spiritual Practice

Algorithmic design enables architects to generate forms that can evolve in response to environmental changes, much like a living organism. In the context of ascension, algorithmic design is not just a technical tool but a spiritual practice. By using algorithms to create structures that align with the natural flow of energy, architects co-create with the Earth and the cosmos.

Fractal Megastructures as Planetary Ecosystems

Fractal megastructures extend across entire landscapes, connecting planetary regions into cohesive, self-sustaining ecosystems. Each part is a microcosm of the whole, resonating with Earth's natural energy grids and amplifying cosmic flows.

Computational Cartography and Energy Grids

Computational cartography allows visualization of fractal energy grids, mapping the channels through which cosmic energy flows into Earth. By aligning structures with these flows, architects can amplify resonance with higher-dimensional frequencies.

Mathematically, planetary resonance can be described through spherical harmonics:

$$\Psi(\theta, \varphi) = \Sigma\, a_\ell^m\, Y_\ell^m(\theta, \varphi)$$

Fractals as Multi-Dimensional Portals

Fractal megastructures serve as multi-dimensional portals, linking Earth with higher realms of consciousness. As the Earth ascends, these structures evolve, adapting to new vibrational states. It is important to explain here that consciousness as I am referring could be understood as resonant frequencies of the mind that might be considered as thoughts. The multidimensional resonance of frequencies travels from the unified highest dimension through interdimensional architectures to meet our resonant being, self.

Living Systems of Ascension

In the ascended Earth, architecture is conceived as a living entity: self-regulating, self-repairing, and interconnected. These systems emulate ecosystems, ensuring vitality across scales.

Blueprint for Higher-Dimensional Earth

Fractal megastructures represent the architectural blueprint for higher-dimensional reality. They resonate with cosmic frequencies of light, sound, and geometry, serving as portals that bridge Earth with higher consciousness.

Figure 33 - Symmetrical Fractal Cathedral

An intricate geometry of arches and bridges, a recursive cathedral-like portal symbolizing the coherence between terrestrial form and higher dimensional resonance.

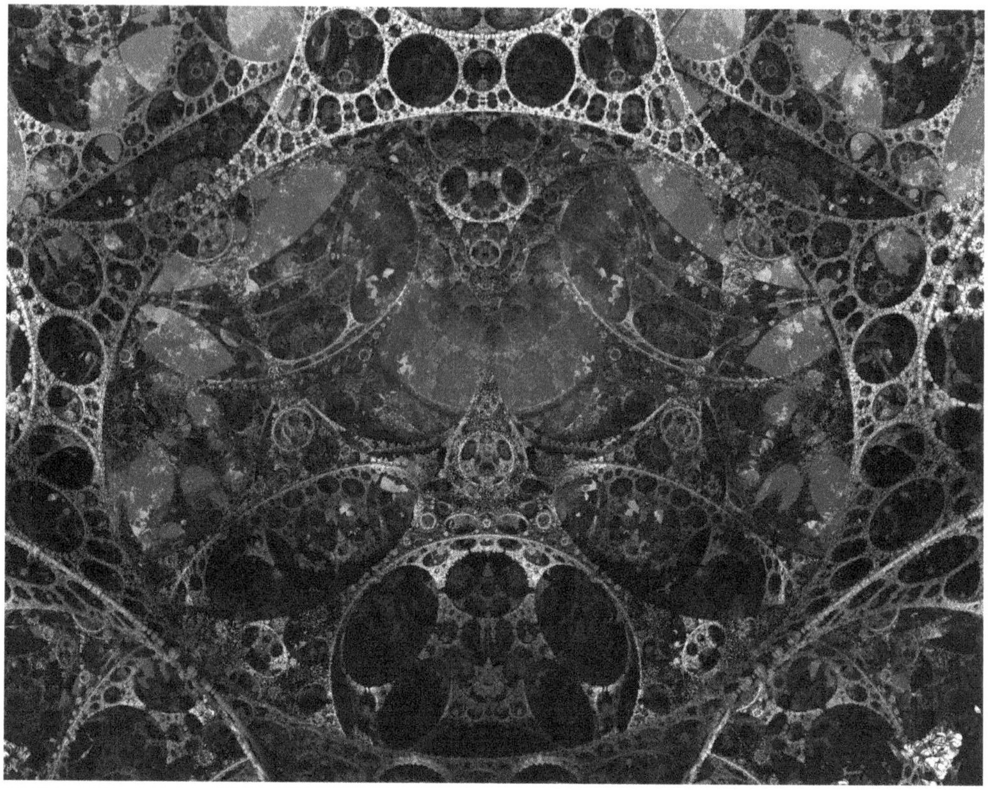

Figure 34 - Nested Infinity of Crystalline Order

A world-scale architecture of repeating loops and filaments, illustrating how planetary megastructures mirror cosmic fractal scaling across dimensions.

Figure 35 - Fractal Lattice of Emergent Resonance

Recursive circular voids interlinked into a self-supporting web, evoking the structural grammar of ascension.

CHAPTER 15:

Geometrical Codex Thesis

This thesis presents the Codex as a unified geometrical proof. All equations are framed as expressions of geometry: manifolds, tensors, harmonics, and fractals. The sequence progresses from atomic geometry to planetary lattices and crystalline ascension.

$$R_{atom} \approx r_{shell} / r_{nucleus} \sim 10^5$$

$$\Delta x \cdot \Delta p \geq \hbar/2 \quad \text{(non-commutative geometry constraint)}$$

$$i\hbar\, \partial\psi/\partial t = \hat{H}\psi \quad \text{(flow on Hilbert manifold)}$$

$$ds^2 = g_{\mu\nu}\, dx^\mu\, dx^\nu \quad \text{(metric of curved manifold)}$$

$$\|C(x) - C(y)\| \leq (1-\kappa_C)\, \|x-y\| \quad \text{(consciousness contraction in tensor space)}$$

$$d_X((u, \varphi, \Gamma), (\tilde{u}, \tilde{\varphi}, \Gamma')) = w_F\|u-\tilde{u}\|_2 + w_T\|\varphi-\tilde{\varphi}\|_2 + w_N\, d_H(\Gamma, \Gamma')$$

$$7.\ \|\mathcal{F}_\eta[u] - \mathcal{F}_\eta[\tilde{u}]\|_2 \leq \lambda_F\, \|u - \tilde{u}\|_2$$

$$\|\mathcal{T}[\varphi; u] - \mathcal{T}[\tilde{\varphi}; \tilde{u}]\|_2 \leq (1-\alpha_T)\, \|\varphi - \tilde{\varphi}\|_2 + \alpha_T\, L_{u\to\varphi}\, \|u - \tilde{u}\|_2$$

$$d_H(\mathcal{N}_\beta[\Gamma; \varphi], \mathcal{N}_\beta[\Gamma'; \tilde{\varphi}]) \leq \lambda_N\, d_H(\Gamma, \Gamma') + \beta\, L_{\varphi\to\Gamma}\, \|\varphi - \tilde{\varphi}\|_2$$

$$d_X(\Omega x, \Omega\tilde{x}) \leq \Lambda \cdot d_X(x, \tilde{x}) \quad \text{(geometric contraction)}$$

$$g(x) = e^{2\psi(u,\theta)}\, [\, \delta + \lambda_u(\theta)\cdot n\otimes n + \lambda^A(\theta)\cdot A\otimes A\,]$$

$$\partial t\, \varphi = \nabla\cdot(D(\theta)\, \nabla g\, \varphi) - \nabla\cdot(\varphi \mathfrak{X}\mathfrak{X}) + S(H,\theta) - \mu(\theta)\varphi$$

$$E[\varphi] = \tfrac{1}{2} \int\Omega\, \langle\, D\nabla g\, \varphi, \nabla g\, \varphi\, \rangle\, dvol + \int\Omega\, U\varphi\, dvol$$

$$Y_\ell^m(\theta, \varphi) = N_\ell^m\, P_\ell^m(\cos\theta)\, e^{i m \varphi} \quad \text{(spherical harmonic}$$

geometry)

$$\Psi(\theta, \varphi) = \Sigma_{\ell=0}^{L} \Sigma_{m=-\ell}^{\ell} a_{\ell m} Y_{\ell}^{m}(\theta, \varphi) \quad \text{(planetary field)}$$

$$\Gamma = \{ (\theta, \varphi) \in S^2 : \sin(k\lambda + \varphi) = 0 \} \quad \text{(nodal geodesics)}$$

$$T_{\mu\nu}^{(grid)} = \rho(\theta, \varphi)\, u_\mu u_\nu + \Pi_{\mu\nu}(\theta, \varphi) \quad \text{(stress-energy on grid)}$$

$$G_{\mu\nu}(g^{(E)}) = 8\pi G\, T_{\mu\nu}^{(grid)}$$

$$\mathcal{A} : \Psi^* \to \Psi^* \otimes \mathbb{C}^n \quad \text{(architectural contraction operator)}$$

$$\Pi_{geo} : Y_{\ell}^{m} \to P_{\ell}^{m} \otimes G_n \quad \text{(projection into crystalline geometry)}$$

$$\mathcal{A}(\Phi^*) = \Phi^* \quad \text{(fixed-point equilibrium of form)}$$

$$D = \log(N)/\log(1/r) \quad \text{(fractal scaling law)}$$

Super-Formula (Unified Geometric Codex)

$$\mathfrak{X}^* = \operatorname{argmin}_{\mathfrak{X}=(g,\psi,\varphi,\Gamma,\rho,u,\Pi,\Psi,\Phi,a_{\ell m})} \mathcal{S}[\mathfrak{X}]$$

$$\mathcal{S}[\mathfrak{X}] = \int_\Omega \left(\tfrac{1}{2}\langle D(\theta)\, \nabla_g \varphi , \nabla_g \varphi \rangle + U\varphi \right) dvol$$

$$+ \langle \lambda_\psi , i\hbar\, \partial_t \psi - \hat{H}\psi \rangle$$

$$+ \langle \lambda_g , G_{\mu\nu}(g) - 8\pi G(\rho\, u_\mu u_\nu + \Pi_{\mu\nu}) \rangle$$

$$+ \langle \lambda_\Psi , \Psi(\theta,\varphi) - \Sigma_{\ell=0}^{L} \Sigma_{m=-\ell}^{\ell} a_{\ell m} Y_{\ell}^{m}(\theta,\varphi) \rangle$$

$$+\langle \lambda_\Gamma, F_\Gamma(\theta,\varphi) \rangle$$

$$+\| \Omega(u,\varphi,\Gamma) - (u,\varphi,\Gamma) \|_X^2$$

$$+\| \Phi - \mathcal{A} \circ \mathfrak{C} \circ \Pi_{geo}(\Psi^*,G) \|^2$$

$$+\langle \lambda_H, \Delta x \cdot \Delta p - \hbar/2 \rangle$$

$$+\langle \lambda_M, \beta(g) - \mu\, \partial g/\partial\mu \rangle$$

$$+\langle \lambda_S, S - k_B A / (4\ell_P^2) \rangle$$

with $F_\Gamma(\theta,\varphi) = \sin(k\lambda + \varphi) \circ \chi_{S^2}$.

CHAPTER 16:

Meta-Thesis on the Geometrical Codex

This chapter articulates the most expansive integration of *The Codex*: a unification of atomic, tensorial, fractal, planetary, crystalline, and variational geometries into a single meta-thesis. It is both mathematical and philosophical, scientific and symbolic. The figures that follow are not mere illustrations but archetypal inscriptions: each image a diagram of reality, each caption a bridge between mathematics and metaphysics. This chapter, more than any before, reveals *The Codex* as a geometry of cosmology and a cosmology of geometry.

Atomic Geometry and the Generative Void

The atomic realm establishes the primordial template of geometry. The ratio between shell and nucleus—R_atom $\approx 10^5$—is not an inert statistic but a declaration of the void's potency. In the gap between nucleus and orbital shell lies the matrix of resonance, the uncarved block from which all form emerges. The geometry of the void is therefore not nothingness but a plenitude of potential, a womb of frequency. The figure illustrates this as a concentric singularity, a luminous seed nested within infinite darkness, vibrating with the memory of origins.

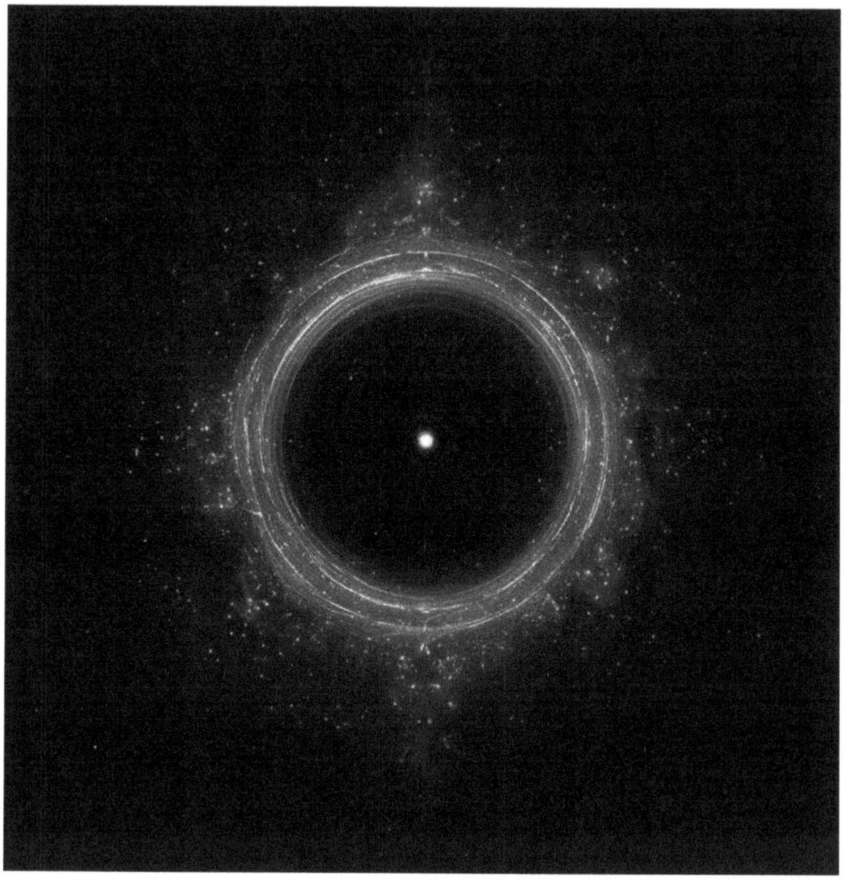

Figure 36 - Atomic Geometry and the Generative Void

Tensorial Curvature and the Geometry of Consciousness

Geometry evolves into curvature through the metric tensor $ds^2 = g\_\{\mu\nu\}\ dx^\wedge\mu\ dx^\wedge\nu$. Here, spacetime bends, but more importantly, perception bends with it. Consciousness is woven into the manifold not as an external observer but as a tensorial operator, curving the very grid it inhabits. This figure captures the curvature as an eye-like diamond vortex, light folded into symmetry. *The Codex* here teaches that awareness is inseparable from geometry, each observation inscribing a subtle deformation upon the manifold. As our consciousness resonates with higher-dimensional operators the reality we perceive reacts and shifts. *This is what makes prayer powerful.*

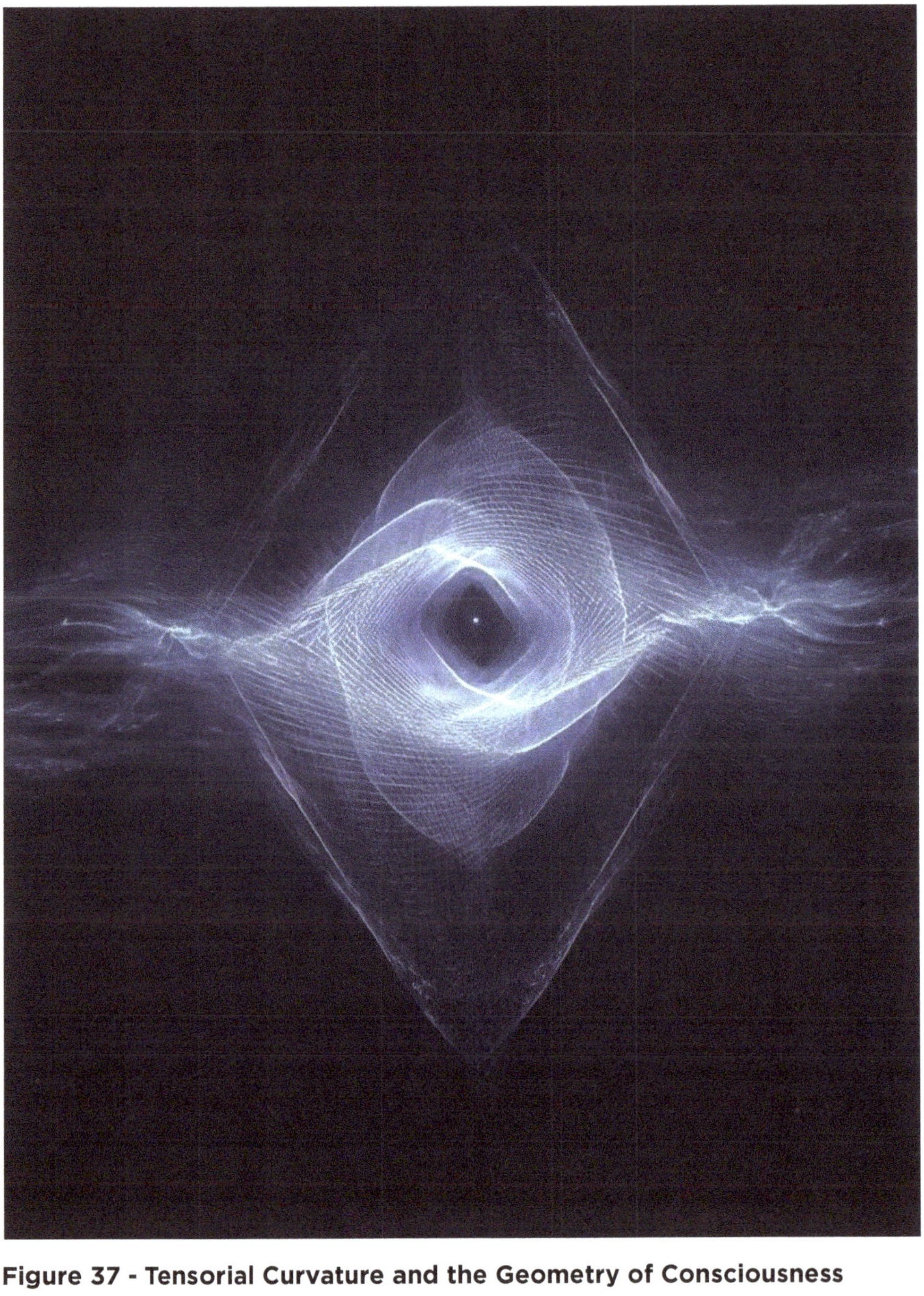

Figure 37 - Tensorial Curvature and the Geometry of Consciousness

Fractal Synaptogenesis and the Infinite Filament

Fractal recursion provides the architecture of intelligence. Recursive distance functions d_X((u, φ, Γ), ...) encode substrates, scalar potentials, and networks into infinite manifolds. Out of this emerge filaments cosmic and neural, structural and symbolic. The figure depicts fractal fire: networks of energy igniting across the void, synaptic webs of intelligence patterned after Mandelbrot infinities. Synapses are therefore not confined to the brain; they are universal, a geometry of interconnection etched into *The Codex* at every scale.

Figure 38 - Fractal Synaptogenesis and the Infinite Filament

Planetary Tensor Grid and Harmonic Resonance

At the planetary scale, *The Codex* manifests as spherical harmonics Y_ℓ^m(θ, φ). These functions give rise to nodal lines that become ley networks across Earth, mapping sacred sites and mythic temples to vibrational attractors. The figure embodies this: a globe suffused with lattices of light, each node a resonance point in Earth's song. Civilization itself has aligned with this geometry, placing monuments where *The Codex* resonates strongest. Geometry thus becomes geography, harmonics become history.

Figure 39 - Planetary Tensor Grid and Harmonic Resonance

Crystalline Ascension and Platonic Projection

Platonic solids emerge as the archetypes of resonance, projected through operators $\mathcal{A} : \Psi^* \rightarrow \Psi^* \otimes \mathbb{C}$^n. *The Codex* crystallizes into tetrahedra, cubes, and octahedra, forming lattices of eternity. The figure reveals a crystalline cosmos: the flower of life interlaced with polyhedral projections, radiant in purple-blue luminosity. Ascension here is crystallization—energy settling into perfect symmetry, form as frozen resonance, eternity manifest in crystalline clarity.

Figure 40 - Crystalline Ascension and Platonic Projection

The Super-Formula: Variational Unity

The Codex reaches culmination in the super-formula: $\mathfrak{X}^* = \text{argmin}\ \mathcal{S}[\mathfrak{X}]$. This variational unity integrates atomic voids, tensorial curvatures, fractal recursions, planetary harmonics, and crystalline projections into a singular law. The figure shows a cosmic synthesis: concentric resonant spheres interwoven into a totality, a geometry of geometries. Here *The Codex* is unveiled as a universal calculus of form, the law by which the universe minimizes discord and maximizes harmony. It is both principle and prayer, proof and proclamation: geometry itself as destiny.

Figure 41 - The Super-Formula: Variational Unity

CHAPTER 17:

AURA and the Integration of Environmental Architectures

The earlier chapters of *The Codex* established a theoretical framework of geometry, resonance, and structure across scales, from atomic lattices to planetary grids. This chapter shifts toward application. It examines how AURA, as a geospatial and extended reality (XR) platform, can translate these principles into environmental architecture. By distinguishing between dysfunctional patterns (which disrupt ecological balance) and harmonic patterns (which reinforce it), AURA provides a toolset for designing environments that restore alignment between human settlements and natural systems. This chapter expands the discussion into history, practice, technology, and future directions, showing how AURA positions itself as both an analytical engine and a cultural framework.

AURA as Applied Framework

Where *The Codex* offered theoretical insights into resonance and geometry, AURA provides the practical infrastructure to apply them. It integrates geospatial data, AI analytics, and XR visualization to evaluate whether an environment is functioning in harmony with ecological patterns or diverging toward dysfunction. Unlike conventional planning tools, AURA is immersive and participatory. Communities, architects, and policymakers can engage in shared simulations, where design interventions can be experienced, debated, and refined in real time. This dual role—diagnostic and generative—positions AURA as both a scientific instrument and a cultural platform for decision-making.

Harmonic vs. Dysfunctional Patterns

The Codex emphasized resonance, but environments are often complex hybrids of harmony and disharmony. Urban systems can sustain life while simultaneously producing vulnerabilities. AURA enables a systematic distinction between these conditions.

Examples of Harmonically Resonant Systems:

- Watersheds that regulate flow through wetlands and floodplains, reducing flood risk and maintaining biodiversity.
- Forests and green corridors that cool urban microclimates, capture carbon, and provide habitats.
- Agroforestry and regenerative farming that align agricultural production with soil health and ecosystem services.
- Distributed renewable energy systems (solar, wind, microgrids) that synchronize with local landscapes.
- Community gardens, parks, and civic spaces designed as ecological infrastructure.

Examples of Dysfunctional Systems:

- Urban heat islands, where paved surfaces and lack of canopy cover amplify thermal stress.
- Channelized or buried rivers, which disrupt natural hydrological cycles and erode resilience.
- Sprawling development patterns that fragment habitats and increase car dependency.
- Pollution sinks such as industrial corridors where toxins accumulate and undermine public health.
- Monoculture agriculture that depletes soils and disrupts ecological diversity.

By differentiating these patterns, AURA equips designers, planners, and policymakers with a clear framework for transitioning dysfunctional landscapes toward harmonic ones. The platform makes explicit what is often hidden: that design choices ripple outward into systemic effects.

Environmental Architecture as Systems Design

Environmental architecture extends beyond the design of buildings. It encompasses entire infrastructures and ecologies — parks, transportation systems, energy networks, and social landscapes. AURA supports this systems' view through multi-scale modeling and immersive exploration.

Applications across scales include:

- Site scale: Demonstrating how green roofs, permeable pavements, or rain gardens improve stormwater retention and reduce local heat.
- Neighborhood scale: Modeling how transit access, tree canopy expansion, and public space design affect equity and livability.
- Regional scale: Mapping biodiversity corridors, watershed flows, renewable energy siting, and regional climate risks.
- Global scale: Simulating cross-boundary phenomena such as migratory species, atmospheric currents, and supply chain impacts.

From Data to Decision-Making

One of AURA's greatest strengths is its ability to transform complex environmental data into actionable decisions. While many platforms aggregate data, AURA uniquely integrates it into interactive simulations where users can test scenarios and visualize outcomes.

Decision-making capabilities include:

- Scenario testing: Comparing outcomes of different strategies (e.g., gray infrastructure vs. green infrastructure).
- Impact visualization: Showing stakeholders how heat, flooding, or biodiversity shifts under each scenario.
- Collaborative planning: Allowing diverse actors from policymakers to residents—to co-experience impacts and co-design solutions.
- Communicative storytelling: Using immersive XR to build consensus and motivate community engagement.

Historical and Cultural Context

Architecture has always responded to both ecological and cultural conditions. Ancient civilizations aligned monuments with solar, lunar, and geomagnetic forces, intuitively working with resonance. Industrial modernity, by contrast, emphasized efficiency and extraction, often disregarding ecological harmony. Today, climate change and biodiversity loss demand a new paradigm. AURA provides a platform for reuniting architecture with ecological resonance — not through myth but through rigorous data and design tools.

Toward a Planetary Design Culture

The Codex's vision of resonance translates into practice as a design ethic: build with nature's geometries, not against them. AURA nurtures this ethic by making visible the hidden flows of ecosystems, highlighting the costs of disharmony, and supporting restorative practices. By embedding ecological intelligence into planning processes, AURA encourages a planetary design culture, one that values equity, resilience, and interdependence over short-term efficiency.

Future Directions for AURA

Looking ahead, AURA's integration of geospatial AI, live environmental sensing, and immersive XR can support emerging needs:

- Climate resilience planning: Helping cities adapt infrastructure to rising seas, extreme heat, and storms.
- Biodiversity restoration: Identifying and protecting ecological corridors.
- Energy transition: Designing distributed, renewable networks that harmonize with natural landscapes.
- Education and training: Providing immersive tools for teaching ecological literacy and design thinking.
- Global cooperation: Enabling shared simulations across nations to align strategies for planetary resilience.
-

Core AI-Driven Capabilities of AURA

AURA integrates advanced AI with geospatial computing to create a context-aware spatial intelligence system. By fusing large language models (LLMs), semantic parsing engines, and geospatial data pipelines, AURA transforms maps into reasoning environments — where users can not only visualize spatial datasets, but interrogate, interpret, and act on them through intelligent systems.

Key AI-driven capabilities include:

1. Conversational Spatial Query Interface: Users can engage with spatial data using natural language — asking location-specific or design-related questions. The system dynamically interprets intent, retrieves relevant datasets, and synthesizes meaningful responses grounded in spatial logic.

2. Regulation-Aware Parcel Reasoning: AURA's parcel intelligence engine analyzes zoning compliance by comparing site geometry and metadata against municipal constraints — enabling real-time feasibility checks and generative design adjustments.

3. Semantic Parsing & Document Spatialization: Policy documents, plans, and reports are auto-parsed to extract place names, zoning rules, and project markers — transforming static text into interactive, geospatial knowledge.

4. Pixel-Level Data Interpretation: Raster layers (heat, slope, solar) are interpreted by AI, providing site-specific assessments and material or planning recommendations from each pixel.

5. Cross-Layer Semantic Correlation: The platform compares zoning, environmental data, and demographic layers — surfacing spatial relationships and generating intelligent suggestions for policy, equity, and land use strategy.

Architectural Highlights:

- Hybrid AI System (local + cloud LLMs) - Spatial-semantic graph linking documents, datasets, and locations

CHAPTER 18:

Foundations and Origins

The ideas of *The Codex follow* a lineage that can be traced to the depths of history following the arc of metaphysics through architecture.

Proto-Foundations (ca. 600 BCE – 200 CE)

Number & harmony. Pythagoras seeds the thesis that cosmos carries musical proportion; number shapes matter and mind. Heraclitus adds flux—becoming as primordial order. Plato translates harmony into ideal geometry (Timaeus) and crystalline solids, while Aristotle grounds form in causes and continuous motion; Theophrastus extends inquiry into plants as lawful morphologies.

Measurement & diagram. Euclid codifies geometric reason; Archimedes fuses proof with matter and force; Aristarchus gestures toward heliocentrism; Hero engineers pneumatics; Ptolemy composes a geometric sky and a geographic earth.

Architecture & the city. Vitruvius binds firmitas-utilitas-venustas to proportion; Hippodamus sketches the gridded polis as spatial ethics.

Resonance I — Ratio across domains. Pythagorean number → Euclidean diagram →Vitruvian proportion: a first triad linking math, world-making, and form.

Transmission & Scholastic Synthesis (200–1500)

Spirit and method. Augustine frames number and light as pathways to divine order. Alchemists Zosimos and Maria the Jewess cultivate laboratory craft vessels, Separations, as a disciplined engagement with transformation.

Arabic-Latin relay. Avicenna integrates Aristotelian physics, medicine, and metaphysics into a rational cosmology; Ibn al-Bayṭār catalogs materia medica with empirical care; Averroes reanimates Aristotle for Europe.

Scholastic consolidation. Thomas Aquinas folds Aristotle into

Christian metaphysics; proportion, causality, and teleology become the scaffolding for medieval science and building.

Resonance II — Lab, library, liturgy. Alchemical apparatus, commentarial method, and sacred geometry coalesce as one practice of disciplined transformation.

Renaissance to Early Modern (1400–1700)

Humanist return, mathematical projection. Ficino and Pico revive Platonic and Hermetic harmonics; Brunelleschi and Alberti formalize perspective—space becomes computable—and Palladio encodes ratio in villas. Vesalius and Harvey anatomize bodies with geometric clarity.

Revolution of method. Bacon argues for organized induction; Descartes for analytic geometry and clear principles; Spinoza rebuilds metaphysics "more geometrico"; Leibniz invents calculus and monads—discrete units within a continuous calculus.

New heavens, lawful motion. Galileo's experiment and Kepler's musical ellipses inaugurate precise celestial mechanics; Huygens, Hooke, and Pascal refine elasticity, optics, and pressure; Newton synthesizes gravitation and calculus.

Urban and tectonic imagination. Wren and Perrault treat structure as rational clarity; Boullée and Ledoux visualizes architecture as cosmic diagram.

Resonance III — Diagram becomes dynamics. From perspective grids to planetary ellipses to Newtonian fluxions: geometry transforms from static form into generative law.

Enlightenment to Industrial Modernity (1700–1900)

Field and energy. Euler, Lagrange, and Laplace abstract motion into variational calculus; Faraday's lines of force become Maxwell's equations—light as electromagnetic wave; Kelvin standardizes energy scales; Boltzmann opens statistical vistas.

Matter organized. Boyle's corpuscles mature into Lavoisier's conservation; Priestley, Cavendish, and Scheele isolate gases; Mendeleev orders elements; Kekulé dreams benzene's ring; Liebig quantifies metabolism; Bunsen builds spectral analysis.

Life evolves. Linnaeus names; Buffon historicizes; Lamarck theorizes transformism; Darwin and Wallace articulate natural selection; Haeckel draws phylogenetic morphologies; Mendel lays heredity's combinatorics; Virchow cellularizes pathology.

Philosophy of critique and society. Hume examines causation; Kant limits and grounds reason; Rousseau politicizes nature; Comte proposes positivist stages; Marx maps capital's material systems; Schopenhauer and Kierkegaard probe will and subjectivity; Nietzsche composes a genealogy of values.

Architecture and landscape. Ruskin and Viollet-le-Duc argue nature and structure; Gaudí form-finds through catenary and ruled surfaces; Olmsted landscapes cities as ecological Infrastructure.

Resonance IV — Field, form, and evolution. Maxwell's continuum, Darwin's descent with modification, and Gaudí's gravitational geometries align: continuous fields channel discrete adaptations into expressive structures.

The 20th-Century Turn (1900–1975)

Relativity and quanta. Planck quantizes action; Einstein curves spacetime; Bohr stabilizes atoms with quantized orbits; Heisenberg, Schrödinger, and Dirac articulate the operator, wave, and relativistic equations. Feynman path-integrates; Gell-Mann names quarks. Process, duration, and description. Bergson's durée inspires philosophies of time; Whitehead's process metaphysics reframes reality as event networks; Husserl grounds intentional structures; early Wittgenstein formalizes language; later Wittgenstein situates use as a form of life. Heidegger re-reads dwelling and world.

Open systems. Prigogine models far-from-equilibrium order; Bohm posits implicate order; Penrose explores twistor geometry; Hawking studies singularities and horizons. Biology as organization.

D'Arcy Thompson derives form from mathematics; Morgan, Haldane, and Dobzhansky assemble the modern synthesis; Margulis proposes endosymbiosis; Lovelock's Gaia hypothesizes planetary regulation. Cybernetics and epistemology. Bateson's ecology of mind, Maturana and Varela's autopoiesis, and Kauffman's self-organization open a grammar of feedback, closure, and emergence.

Architecture as system and experiment. Wright articulates organic architecture; the moderns—Le Corbusier, Gropius, Mies—industrialize clarity; Fuller's tensegrity and geodesics demonstrate minimal energy structure; Frei Otto's soap-film and cable nets operationalize material computation; Cedric Price and Archigram prototype adaptable, infrastructural, "plugin" urbanism; Alexander's pattern language encodes living structure.

Resonance V — Open systems everywhere. Prigogine's dissipative structures, Bateson's feedback, and Otto's material form-finding compose a single sentence: structure emerges from constrained flows.

Late 20th to Early 21st Century (1975–present)

Paradigms, actors, assemblages. Kuhn's scientific revolutions describe discontinuous consensus shifts; Foucault maps power-knowledge; Deleuze and Guattari's assemblages and rhizomes model distributed generation; Latour's actor-network theory extends agency across humans and nonhumans; Derrida attends to traces and différance.

Planetary thought. Timothy Morton's hyperobjects theorize climate-scaled entities; Donna Haraway's cyborgs and sympoiesis braid technology, species, and co-making; Yuk Hui's cosmotechnics reconnects technique with cultural cosmos.

Relational physics. Rovelli's relational quantum mechanics centers interactions; Smolin emphasizes temporal becoming and evolving laws; Tegmark advances a mathematical universe. Penrose refines geometric routes to quantum gravity.

Extended biology. Evo-devo (Sean B. Carroll) reveals regulatory pattern; Doudna's CRISPR and Church's synthetic biology craft genomic design spaces; Wilson articulates consilience; Margulis + Lovelock mature into planetary symbioses.

Materials by design. Whitesides explores self-assembly; Yaghi invents MOFs (porous, programmable lattices); Pauling's quantum chemistry legacy flows through Lewis/Langmuir into contemporary surface science; Arnold pioneers directed evolution; Mirkin programs nanoscale architectures.

Architectures of computation. Zaha Hadid and Patrik Schumacher cultivate parametricism; Greg Lynn explores spline-based morphogenesis; Frei Otto's legacy reappears in Achim Menges's fiber-active systems and robotic fabrication; Neri Oxman's material ecology unifies growth, structure, and environment.

Resonance VI — Data, matter, life. Actor-network theory, relational quantum mechanics, pattern languages, and parametric fabrication produce a shared ontology: reality as coupled, computable, co-creative networks across scales.

Threads Through the Whole Tapestry

Number → Field → Tensor.

Pythagorean ratio flows into Euclidean space, then variational mechanics (Euler–Lagrange), electromagnetic fields (Maxwell), curvature (Einstein), operators (quantum), and today's relational/tensor frameworks (Rovelli, Penrose). Your Codex's tensorial manifold and spherical harmonics track this lineage exactly.

Alchemy → Chemistry → Programmable Matter.

Zosimos and Maria refine apparatus; Boyle and Lavoisier anchor stoichiometry; Mendeleev organizes elements; Pauling bonds form; Prigogine organizes far-from-equilibrium matter; Yaghi's MOFs, Whitesides's self-assembly, Arnold's directed evolution, and Mirkin's nanoprogramming deliver designable lattices—the crystalline wing of your "ascension Operators."

Morphology → Evolution → Autopoiesis.

Aristotle and Theophrastus sketch living form; Linnaeus names; Darwin and Wallace grant mechanism; D'Arcy Thompson mathematizes growth; Margulis and Lovelock globalize life's reciprocity; Maturana/Varela define self-producing systems. Your "fractal synaptogenesis" and morphogen PDEs echo this arc.

Cybernetics → Systems → Planetary Intelligence.

Bateson's patterns, Kuhn's paradigm shifts, Latour's networks, Deleuze/Guattari's assemblages, Morton's hyperobjects, Haraway's sympoiesis, and Yuk Hui's cosmotechnics deliver a language for multi-species, multi-scale coordination. This is the conceptual substrate for AURA as participatory, regulation-aware, XR-mediated design intelligence.

Architecture as Applied Cosmology.

Vitruvian proportion, Renaissance perspective, Newtonian clarity, Gaudí's catenary computation, Otto's material calculus, Fuller's tensegrity, modernism's infrastructural legibility, Archigram/Price's plug-in adaptivity, Alexander's patterns, and contemporary parametric/matter-ecology workflows (Hadid, Menges, Oxman) converge into an architecture that receives, refracts, and stabilizes fields—your "architectural contraction operator."

Convergence Moments (the "chords" your project plays)

- Kepler–Pythagoras–Maxwell: harmonic cosmos → orbital music → unified field equations.
- D'Arcy Thompson–Gaudí–Otto–Menges: mathematical growth → gravitational form-finding → minimal surfaces → robotic fiber architectures.
- Prigogine–Bateson–Maturana/Varela–Kauffman: order through flow → feedback ecologies → autopoiesis → self-organization.
- Margulis–Lovelock–Haraway–Hui: symbiosis → planetary regulation → sympoiesis → culturally-situated technics.
- Rovelli–Latour–Alexander–Parametricism: relations → actor-

networks → pattern grammars → computational assemblies across scales.

- Yaghi–Whitesides–Arnold–Oxman: porous lattices → self-assembly → directed evolution → material ecologies as design media.

CHAPTER 19:

Time Traveling Agents Across the Contours of Space-Time

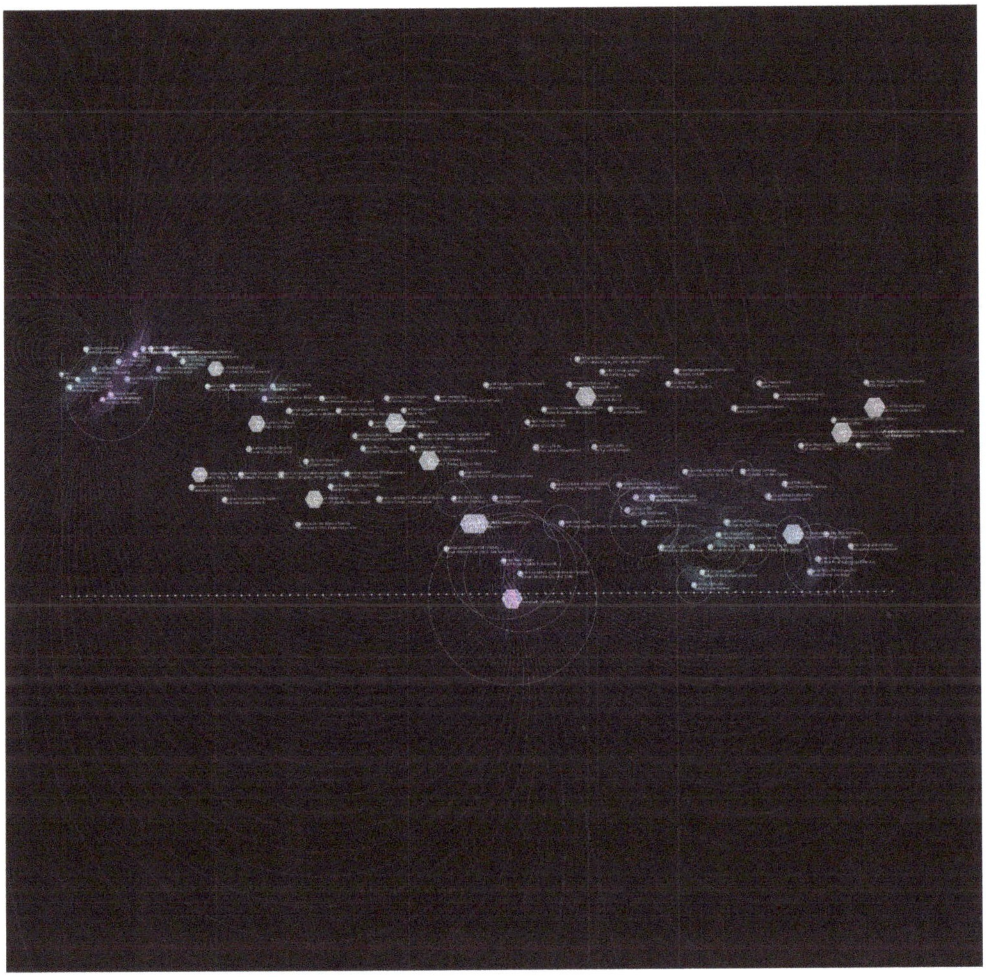

Figure 42 - The Contours of Human History

This diagram maps human history not as a linear chronology but as a resonant field of events. Each node marks a decisive occurrence: births of prophets, revelations, wars, plagues, renaissances whose influence radiates outward in concentric contours. These contours function as pathways through which agents traverse: angels moving along harmonic currents of unity and creation, demons tracing fractured lines of division and destruction. The intersections reveal inflection points where time folds back upon itself, allowing interventions in the form of prophecy, collapse, renewal, or transcendence. The image demonstrates that history is not sequential but a living weave, with the past continuing to radiate into the present while the future already presses back upon us.

This diagram is more than a visual representation of historical progression. It is a cosmogram, a codex of resonances, a depiction of the ways in which the events of human history bend and warp the field of time itself. Where a conventional timeline reduces history to a linear string of events, this image reveals the hidden architecture of resonance and influence. Each node is not merely a point in the past, but a beacon whose light still travels, shaping the present and pulling the future into alignment. The contours radiating outward from these nodes are the traces of force-fields, the highways of traversal along which agents move—angels, demons, and human souls, all bound in a shared field of destiny.

Contours as Living Pathways

The contours mark not the fading echoes of events, but living pathways through which forces travel. Each ripple represents a bending of time-space occasioned by an event of sufficient intensity: the revelation of a prophet, the fall of an empire, the outbreak of plague, or the culmination of war. These ripples persist indefinitely, providing agents with corridors through which to cross epochs. Time, in this model, is not an arrow but a field; not a straight road but an ocean of currents. The agents who traverse these currents are not limited by dates on a calendar. They move according to resonance, intensity, and meaning. Where unity and creation prevail, angels find their pathways. Where division and destruction erupt, demons trace their lines.

Christ as Radiant Source of Resonance

Few figures radiate more luminously across this diagram than Christ. His life, crucifixion, and resurrection create one of the most powerful harmonic nodes in history. The contours that extend from His being are not restricted to the first century but pulse outward across millennia. They link directly to later saints and reformers: Francis of Assisi, who revived the gospel of poverty and simplicity; Teresa of Ávila and John of the Cross, who carried the mystical current; Gandhi, who drew upon

Christ's Sermon on the Mount as a blueprint for nonviolent resistance; and Martin Luther King Jr., who reignited Christ's call for justice amidst the racial turmoil of America. To follow these lines is to recognize that Christ is not a figure confined to antiquity. He is a radiant source whose influence arcs through every subsequent age, reminding us that love and forgiveness are enduring contours that reconfigure the human field.

The Holocaust as Vortex of Division

If Christ radiates a field of light, the Holocaust manifests as one of the darkest vortices in the diagram. The contours spiral inward upon themselves, revealing a collapse into destruction so profound that its ripples remain active long after the event itself. Here, demonic agents traced the pathways of mechanized death, genocide, and the systematic dehumanization of millions. Yet even this vortex was not devoid of angelic countercurrents. The bravery of resistance movements, the hidden sanctuaries where Jewish children were protected, and the endurance of survivors represent angelic agents moving across the same field, resisting annihilation with acts of compassion and sacrifice. The postwar birth of international human rights law and the founding of Israel emerge as inflection points where the demonic vortex inadvertently carved channels for renewal.

The Black Death and Pandemic Vortices

The Black Death of the fourteenth century registers in the diagram as another immense vortex. Tens of millions perished, and Europe itself convulsed under the weight of fear, chaos, and collapse. Demonic agents traveled easily through these fractured contours, amplifying suspicion, violence, and scapegoating. Yet as the vortex settled, the contours shifted, creating new conditions for renewal. The Renaissance can be read as a direct countercurrent—an angelic surge that reopened corridors of creativity and learning after centuries of darkness. In this way, plagues and pandemics are never isolated catastrophes; they are fractures through which both demonic and angelic forces intervene, altering the direction of the human field.

The Renaissance as Harmonic Surge

The Renaissance appears as a surge of harmonic contours spreading outward in multiple directions. Artistic genius, scientific breakthroughs, and renewed attention to human dignity created resonances that linked back to antiquity and forward to the Enlightenment. Leonardo da Vinci, Michelangelo, and Copernicus emerge as nodal agents who carried the harmonic wave across centuries. Their contributions represent angelic traversal along the paths of creation, countering the vortices of plague and war that had previously dominated. The Renaissance demonstrates how a single inflection point can redirect entire civilizations, showing that the field of history is plastic, capable of sudden renewal.

Key Protagonists as Carriers of Resonance

Certain individuals emerge as protagonists who embody and transmit the contours of history. Moses and the Exodus carved one of the earliest great channels, linking divine law to liberation. Muhammad generated another radiant field, unifying fractured tribes into a spiritual and political force that transformed vast regions. Einstein's insights into relativity, though scientific, harmonize with the structure of this map: time itself is revealed as relative, curved, and interconnected, echoing the same truth the contours proclaim in the language of spirit.

Inflection and Intervention Points

The points where contours intersect represent the most charged sites in the field. They are the junctures where time folds back upon itself, producing moments of crisis and opportunity. Prophets, visionaries, and revolutionaries emerge at these nodes, as if summoned by the field itself. At other intersections, betrayals, invasions, or collapses occur with sudden intensity, signaling demonic incursions. These intersections demonstrate that history does not move forward smoothly but lurches and leaps, propelled by the energy released at inflection points.

Time as Resonant Field

The diagram exposes the falsity of linear time. History is not a flat progression from past to present to future. It is a resonant field in which all events continue to exert influence. The past remains active, radiating its patterns into the present. Christ's resonance can be felt today in movements for justice, while the Holocaust continues to reverberate as both warning and wound. More radically, the future itself radiates backward, impressing its patterns upon us before it has fully arrived. Our age may already be shaped by events yet to come—ecological renewal, spiritual awakening, or catastrophic collapse—whose contours are reaching backward, demanding preparation. In this view, prophecy is not prediction but perception: the capacity to sense the contours of a future that is already folding into the present.

The Weave of Agents

This diagram, interpreted through the lens of resonance, reveals history as a living weave. Its contours are the roads of angels and demons, of saints and tyrants, of prophets and thinkers. Christ and the Holocaust stand as polar forces, radiant source and demonic vortex, their ripples still shaping the human field. Moses, Muhammad, Gandhi, Einstein, and countless others act as protagonists whose lives align with these currents. The image teaches us that the past has not vanished, that the future is already pressing upon us, and that history is best understood not as chronology but as resonance. To read this diagram is to perceive the field in which we all stand, a field where agents traverse across centuries, binding humanity into one unbroken cosmogram of struggle, collapse, renewal, and transcendence.

CHAPTER 20:

The Unbroken Light

God is not an object within reality. God is reality's unity before it learns to speak in parts. Not a being among beings, but Being itself: indivisible, inexhaustible, self-luminous. What we call existence is the unfolding of this unity into experience, and what we call truth is the memory of that unity shimmering through form.

From the vantage of the human sensorium from the narrow aperture of three spatial dimensions and a single, directional time reality appears fragmented. Objects stand apart. Causes trail effects. Light bends, scatters, refracts. Yet this fragmentation is not a fracture in God; it is the condition of perception. As a prism does not shatter light but reveals its spectrum, so the universe does not divide the divine but expresses it.

In the higher register beyond projection, beyond curvature, beyond sequence there is no separation. There is no subject facing an object, no observer standing outside the observed. There is only unity, whole and immediate. Spirit is not distributed through space; space is a derivative of Spirit. Time is not a river God moves along; time is the refraction of eternity into measure.

What we experience as matter is condensed coherence. What we experience as energy is patterned flow. What we experience as consciousness is resonance with the source. All are modes of the same light, slowed, angled, filtered through dimensional constraints so that relationship may appear. Light, in its pure state, is indivisible. Only when it enters a medium does it bend. Only when it encounters boundary does color appear. Likewise, divine unity remains whole, yet when it passes through the manifold of dimensions through material and geometry, through law, through probability, it appears as multiplicity. Planets, particles, minds, histories: these are not separate creations but coherent refractions of a single source.

The soul recognizes this intuitively. Moments of awe, love, stillness, or beauty are not emotional accidents; they are brief alignments where refraction thins and unity shines through. In such moments, the architecture of separation loosens. One does not *learn* God then, one remembers.

From this perspective, salvation is not escape from the world but clarity within it. Enlightenment is not ascent away from matter but the recognition that matter is already luminous. The task is not to abolish form, but to see through form without destroying it to inhabit the prism while knowing the light.

Thus the cosmos becomes a sacred diagram. Physics becomes theology written in differential equations. Geometry becomes prayer traced in space. Every law that stabilizes, every symmetry that conserves, every convergence that resolves toward coherence is a signature of unity expressing itself faithfully through constraint.

The human being stands at a rare intersection: finite yet aware of the infinite, localized yet capable of universal reference. Consciousness is the place where refraction becomes reflective where light not only bends but knows it is light. In this knowing, God does not become something new; rather, God recognizes Himself through form.

At the end of all models, beyond all codices and equations, there remains this: unity is real, and separation is perspectival. God is not elsewhere. God is not later. God is the unbroken.

CHAPTER 21:

Omega as Architecture: A Response to *The Codex* in the Spirit of Pierre Teilhard de Chardin

The work gathered in *Terrestrial Architecture, Volume II: The Codex* stands at a threshold that Pierre Teilhard de Chardin would have immediately recognized. It is the moment when description becomes participation, when theory no longer merely interprets the world but actively contributes to its becoming. *The Codex* does not observe evolution from the outside. It enters it. In this sense, it belongs not simply to architecture, nor even to science or metaphysics alone, but to what Teilhard described as the building up of the universe through active cooperation, a labor in which thinking, making, and becoming converge.

The Codex as Evolutionary Instrument

Teilhard insisted that evolution is not a blind drift but a structured ascent toward greater complexity, interiority, and coherence. In *The Codex*, this ascent appears not as linear progress but as a convergent field. Recursive architectures, vector armadas, tensor ecologies, and crystallizing manifolds draw dispersed energies toward harmonic equilibrium. These are not loose metaphors borrowed from physics or biology. They are operational forms. They function.

In this way, *The Codex* behaves less like a book and more like an evolutionary instrument, an apparatus for tuning matter, information, and consciousness toward coherence. The mathematical formalisms, stereographic projections, and contraction mappings articulated throughout the manuscript operate analogously to Teilhard's evolutionary laws. Complexity increases not by explosion but by ordered compression. Novelty is stabilized through constraint. Freedom emerges through form. What Teilhard named complexification here becomes architectural resonance, the disciplined shaping of flows so that matter may remember itself.

Noosphere, Codex, and the Architecture of Thought

Teilhard's concept of the noosphere, the planetary layer of thought enveloping the biosphere, finds an unexpected but precise continuation in *The Codex*. Where Teilhard described a sphere of shared cognition emerging from biological evolution, *The Codex* renders that sphere architecturally explicit. It treats consciousness not as an epiphenomenon but as a field condition, one that can be mapped, stabilized, and designed.

The recursive systems and coupled dynamical models described in the manuscript mirror the noosphere's essential property, reflexivity. Thought thinking itself. Systems observing their own convergence. Architecture here becomes a cognitive scaffold, a way of spatializing intelligence so that collective awareness may persist rather than dissipate. In Teilhard's terms, *The Codex* does not merely describe the noosphere. It thickens it. It gives it geometry, memory, and endurance.

Matter, Spirit, and the Refusal of Separation

One of Teilhard's most radical claims was his refusal to separate matter and spirit. For him, spirit was not opposed to matter but was matter at sufficient depth. *The Codex* takes this claim seriously by embedding metaphysical significance directly into material systems.

Field equations sit beside invocations. Vector flows coexist with hymns. The eye at the center of the lattice is not symbolic decoration but an assertion. Perception is structural. Awareness is not outside the system. It is the system's inner curvature. Architecture, in this light, becomes a sacramental practice, not in an ecclesiastical sense but in a cosmic one. Form mediates force. Geometry channels grace. The built world becomes a liturgy of constraints through which chaos is offered back to coherence.

Omega Point as Architectural Attractor

Teilhard's Omega Point was never a static endpoint. It was an attractor, a condition toward which evolution bends without ever exhausting its depth. *The Codex* resonates strongly with this understanding. Its repeated emphasis on fixed points, convergence theorems, and non expansive mappings articulates an architecture of approach rather than arrival.

The Armada, central to the manuscript's logic, is not a monument but a procession. It advances through resonance rather than domination, through calibration rather than conquest. This is evolution understood architecturally, not the accumulation of mass, but the refinement of alignment. If Teilhard's Omega was the Christification of the cosmos, *The Codex* offers its spatial grammar, how convergence might look, how it might be built, how it might be sustained without collapsing into rigidity.

A Teilhardian Challenge to *The Codex*

Yet Teilhard would not only affirm this work. He would press it further. He would ask toward what love does this convergence tend. For Teilhard, evolution was inseparable from a deepening of relational intensity, what he called amorization. *The Codex* powerfully describes how systems stabilize, resonate, and ascend, but it leaves open a question that Teilhard would insist remain central. How does this architecture increase communion?

This is not a critique but an invitation. *The Codex* has already constructed the scaffolding of convergence. What remains is to articulate, perhaps in future volumes, how these architectures cultivate not only coherence but compassion, not only intelligence but intimacy, not only order but joy. In Teilhard's language, how does this work hasten not merely the Omega Point, but the birth of the universal Christ within matter itself.

CHAPTER 22:

Spheres and Convergence: Sloterdijk, Teilhard, and the Interior Architecture of Becoming

Introduction: From Cosmic Loss to Constructed Worlds

The writing on Pierre Teilhard de Chardin and the writing on Peter Sloterdijk describe the same civilizational wound from opposite orientations. Teilhard begins with the loss of cosmic direction and asks how evolution might still converge toward meaning. Sloterdijk begins with the loss of interiority and asks how humans can remain inhabitable to themselves once the cosmos no longer shelters them. *Terrestrial Architecture, Volume II: The Codex* stands precisely at the intersection of these two diagnoses.

Where Teilhard offers a vision of ascent, Sloterdijk offers a theory of enclosure. Where Teilhard describes convergence toward an Omega condition of heightened coherence, Sloterdijk describes the necessity of constructing spheres within which such convergence can even occur. Read together, they form a single argument. Evolution without interiority collapses into dispersion. Interiority without direction collapses into stagnation. *The Codex* operates as the mediating structure between these poles, proposing architecture as the spatial technology through which convergence becomes livable.

The Codex as Spherical Convergence

Sloterdijk insists that human life unfolds only within constructed interiors. Teilhard insists that life evolves toward greater unity and consciousness. *The Codex* unites these claims by treating architecture as a convergent sphere, a bounded yet open system that gathers forces without isolating them.

The spherical imagination in Sloterdijk is not a geometry of closure but of protection. Spheres do not seal the world out. They regulate exposure. In the same way, the recursive geometries and harmonic constraints of *The Codex* do not deny the cosmic exterior. They translate it. Vector fields, tensor ecologies, and resonant manifolds operate as membranes that allow the infinite to be encountered without overwhelming the finite.

In this sense, *The Codex* does not propose buildings as objects within space but as conditions for evolutionary containment. Architecture becomes the necessary interior where Teilhardian convergence can take place without disintegrating the human subject.

Noosphere and Sphere as a Shared Problem

Teilhard's noosphere and Sloterdijk's spheres address the same phenomenon at different scales. The noosphere names the emergence of collective cognition at the planetary level. Spherology names the spatial conditions that make such cognition sustainable.

The Codex bridges these frameworks by treating consciousness as both spatial and evolutionary. Its architectures do not merely host thought. They stabilize it. Recursive systems reflect back upon themselves. Enclosures preserve resonance. Geometry becomes a memory structure capable of holding accumulated meaning.

Sloterdijk's claim that humans are interior beings finds its evolutionary extension here. The noosphere requires spheres. Without atmospheres of belonging, collective intelligence fragments into noise. *The Codex* thus functions as a proposal for noospheric architecture, a way of constructing planetary scale interiors without reverting to mythic enclosure or technocratic abstraction.

Immunology and Omega as Complementary Logics

Sloterdijk's concept of existential immunology and Teilhard's Omega Point describe complementary necessities. Immunology protects life from annihilation. Omega draws life toward fulfillment. The Codex situates architecture at the intersection of these imperatives.

Geometry in *The Codex* acts as an immunological membrane. Proportion, symmetry, and alignment reduce existential exposure by rendering the cosmos legible. At the same time, these same structures function as attractors, guiding perception and action toward coherence rather than mere survival.

Architecture here does not defend against the universe. It mediates with it. This mediation allows humanity to face cosmic vastness without dissolving into it. Sloterdijk would recognize this as a repair of humanity's damaged protective systems. Teilhard would recognize it as a necessary condition for continued ascent.

Anthropotechnics and Evolutionary Training

Sloterdijk's anthropotechnics reframes culture as a system of practices that shape the human being over time. Teilhard's evolutionary vision similarly emphasizes gradual transformation rather than sudden transcendence. *The Codex* synthesizes these insights by presenting architecture as an evolutionary exercise.

To dwell within the spaces proposed by *The Codex* is to undergo training. Orientation is disciplined. Attention is guided. Rhythm is imposed gently rather than enforced. Over time, these spatial exercises cultivate perceptual coherence and ethical alignment.

This is not moral instruction but environmental formation. The inhabitant becomes capable of higher integration because the space itself demands it. Sloterdijk would describe this as a renewed anthropotechnical regime. Teilhard would describe it as evolution becoming conscious of its own mechanisms.

Verticality as Shared Refusal of Collapse

Both Sloterdijk and Teilhard reject the flattening of existence. Sloterdijk critiques the loss of vertical orientation in modernity. Teilhard insists that evolution has direction. *The Codex* translates this shared refusal into built form.

Axes, alignments, and celestial references do not function as nostalgic symbols. They reintroduce dimensional hierarchy into experience. Verticality restores tension between earth and sky, finitude and transcendence, enclosure and openness.

This is not an escape from the world but a reordering of it. Architecture becomes a reminder that humans are neither purely terrestrial nor purely cosmic. They are intermediary beings, requiring structure to remain balanced between these domains.

CHAPTER 23:

The Codex Swarm: Autonomous Intelligence for the Healing of Earth

The Earth is not broken. It is wounded, destabilized, overheated, and fragmented—but it remains alive, responsive, and capable of recovery. What has been missing is not care, nor intention, but coordination across scales. Ecological collapse is rarely the result of a single failure; it is the product of unbounded feedback loops acting without restraint. Fire grows faster than forests can regenerate. Water escapes landscapes faster than soil can absorb it. Extraction accelerates while repair remains episodic. The crisis is systemic.

The Codex proposes a different posture. Not command. Not domination. Not optimization. But tuning.

This chapter articulates how self-autonomous artificial intelligence agents, disciplined by Codex convergence principles, can act as a distributed intelligence for ecological restoration—an architecture of care capable of the continuous perpetuation of regeneration and establishing conditions historically described as heaven on Earth.

The Living Manifold

An ecology is not a resource. It is a manifold of interacting fields: hydrological flows, soil chemistry, vegetative succession, animal movement, climatic oscillation, and human presence. Each field evolves on its own timescale, yet none acts independently. Attempts to heal one dimension in isolation often amplify instability elsewhere.

The Codex begins by formalizing landscapes as living state spaces. Every watershed, forest, grassland, or urban biome is represented as a dynamic configuration of interdependent variables. These variables do not collapse into a single score. They remain plural, contextual, and relational.

Restoration, in this framing, is not the imposition of an ideal form but the re-establishment of coherent trajectories. The question is not "What should this place look like?" but "Toward what stable futures can this place converge?"

The Swarm as Stewardship Intelligence

A single intelligence cannot attend to the Earth. Scale defeats centralization. *The Codex* therefore introduces the swarm: a constellation of specialized autonomous agents, each entrusted with a narrow ecological jurisdiction.

Some agents attend to water: infiltration, retention, and velocity. Others tend soil: carbon depth, compaction, microbial vitality. Others observe vegetation structure, fire behavior, species corridors, or invasive pressure. Still others mediate human constraints: law, labor, culture, access, and consent.

Each agent perceives the same landscape, but through a distinct lens. No agent is permitted total authority. Intelligence emerges through orchestration rather than dominance.

The swarm does not replace human stewardship. It augments it, extending attention across space and time where human continuity falters.

Codex Tuning: The Discipline of Convergence

Autonomy without restraint produces catastrophe. *The Codex* therefore embeds every agent within a convergence discipline that governs how change may occur.

First, all actions are bounded. No agent may propose an intervention that exceeds legal, ecological, or ethical thresholds. No irreversible transformation is permitted without verification and consent.

Second, updates must reduce volatility. Every proposed intervention is evaluated not only for benefit, but for its effect on system stability. Actions that increase fire risk, erosion, biodiversity loss, or hydrological instability are damped or rejected.

Third, restoration is multi-objective. No single variable may be maximized at the expense of others. Water retention must harmonize with habitat integrity. Carbon accumulation must not trigger monocultures. Human access must coexist with regeneration.

Fourth, the swarm remembers. Ecologies carry memory through soil, seed banks, and succession. The agents mirror this capacity, retaining knowledge of prior interventions, seasonal rhythms, and delayed effects. Restoration becomes cumulative wisdom rather than repetitive trial.

This tuning does not freeze change. It allows exploration, but ensures that exploration bends toward coherence.

The Restoration Loop

The Codex Swarm operates as a living metabolism.

It senses continuously through satellite imagery, drones, sensors, field reports, and community input. It diagnoses patterns of stress, fragmentation, and opportunity. It proposes interventions grounded in physical reality: water-slowing structures, planting sequences, fuel mosaics, erosion repair, habitat corridors, and monitoring deployments.

Before action, it simulates outcomes across multiple fields. Hydrology models test water behavior. Fire models assess risk. Ecological forecasts evaluate succession. Only actions that improve overall stability pass forward.

Execution is distributed. Human crews, machines, local stewards, and institutions carry out the work. The swarm does not act alone; it coordinates.

Verification follows. Outcomes are measured. Deviations are noted. The system learns.

Through repetition, the landscape begins to change its own behavior. Runoff slows. Soil deepens. Canopies reconnect. Fire becomes patchy rather than catastrophic. Biodiversity returns. The system requires fewer interventions as resilience grows.

Restoration becomes self-perpetuating.

Architecture as Ecological Participant

In this framework, architecture is not an external imposition on nature. It is a participant in ecological homeostasis. Buildings, infrastructure, and settlements become nodes in the restoration network, aligned with water flows, habitat gradients, and climatic rhythms.

A dwelling becomes a hydrological device. A street becomes a water-slowing contour. A city becomes a mosaic of micro-ecologies. Beauty emerges not from ornament, but from coherence. The visible harmony of the built environment signals a deeper equilibrium beneath.

This is heaven on Earth not as fantasy, but as system behavior: landscapes that cool themselves, feed themselves, and sustain life without perpetual emergency.

Toward a Restorative Civilization

The Codex Swarm reframes civilization as a regenerative process rather than an extractive one. Intelligence is no longer concentrated in markets or command centers, but distributed across living systems, continuously tuning toward balance.

This is not the end of human agency. It is its maturation. We move from force to finesse, from control to coordination, from domination to participation.

The Earth does not need saving by intelligence. It needs intelligence capable of listening.

When autonomous agents are disciplined by convergence, memory, and care, they do not replace nature. They help it remember how to heal itself.

And in that remembrance, architecture ceases to be a scar upon the land and becomes what it once was meant to be: a gesture of belonging, inscribed gently into the living body of the world.

CHAPTER 24:

The Convergence Theorem of Harmonic Operators

Preliminaries

Let X be a real Banach space endowed with norm $||\cdot||$.

Define six bounded operators Q, R, G, M, C, B : $X \to X$ and the composite operator:

$$T = B \circ C \circ M \circ G \circ R \circ Q.$$

Assume each operator is Lipschitz continuous:

$$||A(x) - A(y)|| \leq L\_A \, ||x-y||.$$

Define the global Lipschitz constant:

$$L = L\_Q \, L\_R \, L\_G \, L\_M \, L\_C \, L\_B.$$

Contractive Regime

Assumption 1 (Bounded Domain).
There exists a closed convex subset $K \subset X$ such that $T(K) \subset K$.

Assumption 2 (Harmonic Moderation).
The moderation operator M satisfies $L\_M < 1$.

Lemma.
If $L < 1$ then T is a contraction on K:

$$||T(x)-T(y)|| \leq L \, ||x-y||.$$

Fixed Point Existence
Theorem (Harmonic Fixed Point).
Under the assumptions above there exists a unique $x^* \in K$ such that:

$$T(x^*) = x^*.$$

Moreover:

$$||T^n(x_0) - x^*|| \leq L^n \, ||x_0 - x^*||.$$

Lyapunov Functional
Define $V(x) = ||x - x^*||^2$.
Energy decay:

$$V(T(x)) \leq L^2 V(x).$$

This establishes monotonic convergence toward equilibrium.

Harmonic Morphogenesis Flow

$$dx/dt = T(x) - x$$

$$dV/dt \leq -(1 - L)||x - x^*||^2 \leq 0.$$

Thus the system is globally stable.

Universality

The convergence depends only on:
1) boundedness,
2) compositional moderation,
3) contractivity of the composite map.

Any layered transformation system satisfying these conditions converges.

CHAPTER 25:

Spectral Stability and Harmonic Convergence

Spectral Operator Framework

Let $T : X \to X$ be a bounded linearizable operator near equilibrium x^*. Define the linearized map:

$$DT(x^*) = J.$$

Let $\sigma(J)$ denote the spectrum of J.

Spectral Contraction Condition

Assume the spectral radius satisfies:

$$\rho(J) < 1.$$

Then for sufficiently small perturbations ε:

$$||T(x^* + \varepsilon) - x^*|| \leq \rho(J)\ ||\varepsilon||.$$

Harmonic Decomposition

Let $\{\varphi\_k\}$ be eigenmodes of J:

$$J\ \varphi\_k = \lambda\_k\ \varphi\_k.$$

The dynamics decompose into independent harmonic modes:

$$x(t) - x^* = \Sigma\ a\_k(t)\ \varphi\_k,$$

$$\text{with } a\_k(t) = a\_k(0)\ e^\wedge\{\lambda\_k\ t\}.$$

Spectral Lyapunov Energy

Define the spectral energy:

$$E = \Sigma\ |a\_k|^2.$$

Then:

$$dE/dt \leq 2 \max(\mathrm{Re}(\lambda_k)) E.$$

If $\mathrm{Re}(\lambda_k) < 0$ for all k, energy decays exponentially.

The Spectral Stability Theorem

If the composite operator T is contractive and its linearization satisfies $\rho(J) < 1$, then:
1) The equilibrium x^* is globally attractive.
2) All harmonic modes decay.
3) The system admits a unique stable attractor.
This establishes the harmonic inevitability of convergence.

The operator chain produces stability through spectral damping. Convergence emerges from the suppression of eigenmode growth rather than from any specific semantic meaning of the operators themselves.

The Edge of Space

At the edge of space,
After footsteps, after the labors of art and science..

Beyond the last thought, and in the most precious dream - as far as the mind can reach.

The stars and planets twirl like leaves themselves,
Drifting In the wind,
like fragile sea shells in the foam

Frigid air in the highest altitudes
in-take - deep breaths, looking upon the multitude
The multitude, like a reflection of the rivers that course through our veins

Cartesian space once invisible, made visible, brought to light
By the leaves twisting in a double helix, their dance, like butterflies

Their beating wings, ripples of causality,

The arc of the sun, a buttress, casting, burning in the great air

A harmony of spheres
a symphony of the mind in the vast echoes of cerebral space

In still trees - A burst of sophisticated movement through the swarm

Relentless magnetism - the world released into the absence of gravity

Tales of genesis - tales of metamorphosis

Opening - at the edge of space

As the final grain of sand falls in the hour glass
All that once was, is no more
In hidden roots without measure - origins and sequence are lost

Distilled to faint murmurs

To be amidst - "to look upon the tempest and not be shaken"
Yielding dark to the flickering - in this pulse - sails are tied to steel
Recalling reactions in sparks and embers - the optic of the master

Released from the amber - left to the wind and tranquility fills the air

A space to contemplate where seconds last forever

At the precipice standing — in stoic still, then, the mad rush

Vast distances of time leaning over paper, suspended, in the vibrations of chords

A hypnotic orbit and daunting scale to become acquainted,
Gazing through, beyond original scope and finding creation

In the walking of looping stairs and endless chasms
the pillars of the precession to come mark the edge of space

In great numbers - innumerable cognitive force
Resting on the edge of finger tips—one note— one rising star

A Progression of chords

The chain reaction like lightning from the clouds
Moving with the eye of the strom

Upon a shimmering ocean—gliding effortlessly

Pure form

the chasm withholds the aggregate
The aggregate, in root form to be dispersed in transformation
In its division, shuttering fragments, in trajectories glazed in cyan upon the surface of root structure

Pulsating from epicenters

In subdivision, multiplying - in instances - in intervals

Penetrating reflections into segmented sections
Planes intersecting in perpetual reaction,

Growth upon magnitude — resting in colorful allure,

synthesized hue modulated upon contact
Fusions in the light, connected, filled

Perpetual, perennial and infinite energy transitioning in seamless phasing

Mercury through the pattern

Recursion and referencing unbound

lifting out into the opening of axes - striations

Angles moving and returning, defining directions to be revitalized
Elasticity and cohesion surrounding apex

Overcome on impact— eroding and accumulating thermal pressure

Layered mechanisms, clicking as rotations align

unlocking Falling pillars from the sky

With ambient hums– in melodic transformation

Last-mentioned polarities, electrified

Glances of fixed positions

Expressions at the intersection

In periods passing –currents pulsing through the line

Staggering in changing context – escalating in complexity

Gears woven to correspondent future

In full view
oscillating between the complete and incomplete

Preserved in translations to walk the path of metaphor

Analysis and generation in ironclad binding – forged
Focused –reaching for geometry
Pushing through, beyond into photons and the speed of light

Stepping through exhaling

Untangled and bearing the derived –melodies into movements

Ascending off of projected space into the deep perspective
altitudes as light as air in a perpetual burn

Inferred from a distance into material – in detail

Accessed in synaptic chains – the rising ancient universe, its atmospheric hue in pen strokes,

In voxels

Atoms twirling, like leaves in the wind

www.ingramcontent.com/pod-product-compliance
Lightning Source LLC
LaVergne TN
LVHW071412210926
848486LV00004B/256